Natural Farming

NIPA® GENX ELECTRONIC RESOURCES & SOLUTIONS P. LTD.
New Delhi-110 034

About the Author

Dr. Rahul Dev Behera holds a Ph.D. in Soil Science and Agricultural Chemistry from Odisha University of Agriculture and Technology (OUAT), Bhubaneswar, completed in 2021. He is currently serving as a Subject Matter Specialist (Soil Science) at Krishi Vigyan Kendra (KVK), Bargarh, under OUAT, Odisha.

Dr. Behera has made notable contributions to agricultural research and extension. He has published 22 research papers in reputed national and international journals and authored 4 books along with 10 book chapters. His work has significantly impacted rural agriculture in the Malkangiri and Bolangir districts of Odisha, where he has effectively transferred various innovative technologies to farmers.

As part of his efforts to promote soil health, he has tested over 2000 soil samples at the KVK Soil Testing Laboratory and facilitated the distribution of more than 5000 Soil Health Cards. Through numerous skill development and technology dissemination trainings, he has empowered farmers to adopt modern practices, leading to enhanced productivity across different farming enterprises.

Dr. Behera has been the recipient of several prestigious accolades:

Young Scientist Award in Agricultural Science, International Multidisciplinary Research Foundation, 2017

Fellowship Award in Environmental Science, Environment & Social Welfare Society, 2017

Dr. A.P.J. Abdul Kalam Award with Medal in Soil Science, Odisha Rajya Sanskruti Surakhya Mancha, 2017

In addition to his field and research work, he engages in farmer education through radio talks, newspaper articles, booklets, and leaflets. Currently, Dr. Behera is actively promoting Zero Budget Natural Farming (ZBNF) as a sustainable alternative to counter the degradation of soil health due to excessive chemical fertilizer use. He regularly conducts training, awareness, and demonstration programmes on natural farming practices across various cropping systems.

Natural Farming Principles and Practices

Rahul Dev Behera
SMS(Soil Science)
Krishi Vigyan Kendra, Bargarh
Odisha University of Agricultural Technology
Bhubaneswar, Odisha

NIPA® GENX ELECTRONIC RESOURCES & SOLUTIONS P. LTD.
New Delhi-110 034

NIPA® GENX ELECTRONIC RESOURCES & SOLUTIONS P. LTD.

101,103, Vikas Surya Plaza, CU Block
L.S.C.Market, Pitam Pura, New Delhi-110 034
Ph : +91 11 4386 0225, 9717133558, 9540816132
E-mail: newindiapublishingagency@gmail.com
Website: www.nipareources.com

Print ISBN: 978-93-58877-50-2

ebook ISBN: 978-93-58875-65-2

Composed and Designed by NIPA®.

Preface

In an era defined by accelerating technological progress and the intensification of industrial agriculture, we face mounting challenges that threaten the very foundations of our food systems. Soil erosion, declining biodiversity, water pollution, and the loss of traditional ecological knowledge have become global concerns, prompting farmers, researchers, and policymakers to question whether modern practices are sustainable in the long term. This book emerges from a conviction that, while innovation is critical, looking to nature itself offers invaluable guidance. Natural farming—a holistic approach that respects the integrity of ecosystems—provides an alternative framework for cultivating food, nurturing soil health, and fostering resilient communities.

Natural farming does not merely reject chemical inputs or mechanized methods; it embodies a philosophy of harmony with ecological processes. Pioneers such as Masanobu Fukuoka in Japan and Subhash Palekar in India have demonstrated that working in concert with natural cycles—rather than imposing human will upon them—yields healthier soils, more robust yields, and a closer relationship between people and the land. The methods advocated by these practitioners often involve minimal tillage, a reliance on locally available resources, the encouragement of biodiversity through intercropping and cover crops, and the avoidance of synthetic fertilizers or pesticides. In essence, natural farming invites us to observe, listen, and learn from the web of life that sustains us.

This preface introduces readers to the principles and promise of natural farming. Chapter by chapter, we will trace the evolution of its core tenets: embracing "do-nothing" techniques that allow nature to self-regulate, fostering symbiotic relationships among plants and microorganisms, and creating systems that require less external input while delivering long-term benefits. Alongside theoretical discussions, we explore real-world case studies from diverse regions—rice paddies in Kerala, fruit orchards in California, and vegetable plots in the English countryside—each illustrating how natural farming can be adapted to varied climates, soils, and cultural contexts.

A fundamental tenet of this approach is the rehabilitation of soil. Modern agriculture often treats soil as an inert growing medium, subject to depletion.

By contrast, natural farming regards soil as a living organism, teeming with microbial life and organic matter. Techniques such as mulching, crop rotations, and the use of compost help to rebuild soil structure and fertility. Over time, farmers who adopt these practices witness a remarkable transformation: greater water retention, reduced erosion, and the resurgence of beneficial insects and earthworms. These improvements not only reduce dependence on external inputs but also foster a sense of stewardship and respect for the land.

Beyond ecological restoration, natural farming has profound social and economic implications. By reducing reliance on purchased chemicals and machinery, smallholder farmers can reclaim autonomy over their livelihoods and reinvest savings into their communities. Furthermore, consuming produce grown through natural methods offers health benefits to families and consumers, as crops contain fewer residues and better nutrient profiles. Throughout this book, we delve into anecdotes and data that highlight how natural farming can uplift rural economies, strengthen food sovereignty, and inspire a new generation of farmers to embrace regenerative practices.

Author

Contents

1

Introduction

In history, some developments were possible through belief and scientific study with evidence to support their claims. Our ancestors made ways of surviving on earth by varying the raw materials and tools. Science had absolutely no place, and all discoveries were derived through the trial-and-error method. The same was for agriculture, tracing back to our roots, the farming system was highly dependent on the on-farm inputs. However, with the timeline of evolution, the exponential growth of the population demanded speedy growth in agricultural production giving rise to Green Revolution technologies (GRTs) in the mid-1960s, which introduced high-yielding varieties of crops that were responsive to higher dosages of chemical fertilizers and irrigation and persuaded farmers to resort to intensive monocropping. Since the onset of GRTs in the mid-1960s, food production in India has risen five to ten-fold (Dasgupta, 1977).

However, while the past few decades have seen substantial 'green revolution' advances, I admit, the hazards have kept themselves hidden and well concealed--of these are significant adverse impacts on the natural resources, agriculture and its produce and indeed extend to human health (Koner & Laha, 2021). There is further proof that increasing dosages of chemical fertilizers led to an emergence of pest resurgence; pesticides residues considerably exceeding permissible limits contaminated drinking water and/or air often, a drop in biodiversity, nitrogen leaching and groundwater pollution, and soil heavy-metals build-up are very common in intensive agriculture regions (John & Babu, 2021). While such issues contributed to suicides in rural India, their unwarranted use also diminished pollinator populations (Baron et al., 2017). There are many measures to ease the overwhelming cost burden upon farmers on account of farming and rising costs of cultivation. It pertains to the increased input subsidies made by various state and central governments-the fertilizer subsidy, the free supply of electricity for irrigation, the interest subvention on agricultural credit, and the premium subsidy for crop insurance. On the contrary, the recent trends show that 75% constitutes a higher number of electricity subsidies, reserved for agriculture consumers, from 2016-2019.

The direct subsidy alone in 2019 amounted to INR 1103.91 billion and a cross-subsidy of at least INR 750.27 billion (Aggarwal et al., 2022).

Many state governments have provided free electricity, while others offered subsidies of 75-80% for agricultural purposes. Institutional credit to agriculture sectors in 2020-2021 stood at INR 15,754 billion. If there is 3% interest subvention, it indicates that such subsidy would amount to INR 473 billion (Ministry of Agriculture and Farmers' Welfare, 2022). The premium payment by both central and state governments for crop insurance during 2021 was pinned at INR 570 billion (PMFBY, 2022). All these factors appear to make a perfect storm both for small and tenant farmers and the long-term sustainability of Indian agriculture. Thus, the Indian farmers are getting entrapped more and more in this vicious cycle of a debt trap compounding their distress. Natural Farming (NF) is the only one in its kind, referring to chemical-free farming approached as an agroecological approach (Rosset, 2012). The agroecological practice began with one Japanese farmer, Masanobu Fukuoka, whose local customized version was introduced in India in the mid-1990s by Sh. Subhas Palekar, an Indian agriculturist, and called 'Zero Budget Natural Farming (ZBNF)'. The crux of Natural Farming practices is free Jeevamritha and Beejamritha. Jeevamritha is a liquid fermented concoction containing cow dung, cow urine, jaggery, pulse flour, and bunk soil mix with water containing several beneficial microbial colonies; it serves as a bio-stimulant which will reactivate the soil microorganisms in the field and foliar microorganisms when sprayed on leaves. Beejamritha is really Jeevamritha but without water and is used for seed treatment. For this technique, beneficial microbes should inspire the roots and leaves of the germinating seeds, thus enabling healthy growth. Other significant inputs are Achhadana (bio-mulching), intercropping, and sowing seeds of local sources.

Such a farming system restores several self-ready preparations of biologically designed products vis-a-vis insecticides, including neemastra, agniastra, and bramhastra (Mishra, 2018). These preparations check pests like mealy bugs, sucking pests, fruit, stem and pod borers, leaf rollers, etc. The full spectrum of the natural farming regime has partly promoted soil-based features or improvements through quick development of heterotrophic microbial communities and a concurrent increase in soil organic matter (Shyam et al., 2019; Smith et al., 2020; Kumar, 2023 & Saharan, 2023). Some studies have indicated reduced yields (Kumar, 2023; Babalad & Navali, 2021; Naik, 2020), whereas other studies have reported no yield reduction (Kumar et al., 2019). Recently, the government of India has been promoting this system at a very huge level, aiming to encourage chemical-free farming. In a speech to the nation on the 76th Independence Day of India, the Prime Minister of India said, "ZBNF is a promising tool to minimize the dependence of farmers on purchased inputs and it reduces the cost of agriculture through reliance on traditional field-based technologies which, in turn, improve soil health" (Duddigan et al., 2022).

Additionally, there are other schemes as such that can be cited-National Mission on Natural Farming, Paramparagat Krishi Vikas Yojana under the sub-mission of Bharatiya Prakritik Krishi Paddhati, Andhra Pradesh Community Natural Farming, Mission Organic Value Chain Development for North Eastern Regions, etc.-which are popularizing the adoption of natural farming by farmers all over the different parts of the country. The BPKP scheme provides a provision of financial assistance of INR 12, 200/ha (≈147 USD/ha) for 3 years for cluster formation, capacity building, and continuous handholding, certification, and residue analysis. The apex research body, Indian Council of Agricultural Research (ICAR), has undertaken a study on the evaluation of Natural Farming of certain crops (Press Information Bureau, 2922 & 2023). The overall popularity of Natural Farming has drawn the attention of many sections of society. It is estimated that more than 500, 000 hectares of land in India planted to Natural Farming (Economic Survey, 2022) and it's expected to scale up with 14 million hectares of land under Natural Farming by 2025 under the PKVY scheme (Jain, 2022). Scaling up of Natural Farming may not depend alone on agricultural practices; social factors, such as social movements, public policies, markets, pedagogical processes, leadership, and discourse, also play a significant role in the scaling up of Natural Farming (Khadse et al., 2018; Cacho et al., 2018; Khadse & Roset, 2019).

Farmer-focused and farmer-led knowledge exchange form the core basis of the long-term dissemination of natural farming (NF) practice (Bharucha et al., 2020). Against that backdrop, the present study tries to address core questions, such as: In what form is natural farming being practised by Indian farmers? What would happen to the crop yield and income of the farmers? How are natural farming practices helping the agro-ecology? The present study is based on a field survey conducted in three major states of India (Karnataka, Andhra Pradesh, and Maharashtra) believed to have a high adoption of natural farming. The pattern of adoption of different components of NF by the farmers has been looked into, and the crop yield and farm income during NF practices have been compared with other existing farming practices. According to the FAO, overall food production must be increased by 70% by 2050 to feed the rising world population and changing consumption patterns in developing countries. Moreover, during that very time, India will become the most populous nation with 1.51 billion people. Global food supply would become a major issue for such an Indian population in terms of food security for its people. Any farming practice or production technology which has not been scientifically endorsed and/or is likely to adversely affect yield may have far-reaching implications for the national goal of food and nutritional security. The technology of the 'Green Revolution' characterized by the intensive use of high-yielding varieties, land-intensive cropping (which involves the use of chemical fertilizers and irrigation), introduced in the 1960s, helped the nation to over Waspsome food shortages. However, intensive agriculture had led to large environmental degradation, the eutrophication of land and water bodies, GHG emissions along with biodiversity loss (See Evenson and Gollin, 2003; Canfield et al., 2010; Smith et al., 2013; IAASTD, 2009; Pingali, 2012).

On the contrary, natural farming has, dissolved and cleaned, never been linked to any form of agroecology-based diversified farming process, and thus functions as a chemically-based program for varied farming, intermixing crops with trees and livestock with functional biodiversity (LVC, 2010; Rosset and Martinez-Torres, 2012). A ZBNF realized an introduction and dissemination by agriculturist Sh. Subhash Palekar as early as the middle of the 1990s genuinely, and he was rewarded by literally one of the foregone extremely high civilian awards of India, Padma Shri in 2016 for advocating these strategies for alternative farming practices (Khadse et al., 2017; Mishra, 2018; Niyogi, 2018; Economic Survey, 2019). Usually, its implementation covers expansive levels within various states, particularly Andhra Pradesh, Karnataka, Maharashtra, Himachal Pradesh. This is judged to considerably bring down the cost of production since in place of chemical fertilizers and pesticides, it utilizes homemade products such as Jeevamritha, Beejamritha, Neemastra, etc. This has, therefore, been targeted toward a reduction in the consumption of farmyard manure or farmyard waste compared with that intended in organic farming by intercropping and mulching (Palekar, 2005; 2006). Further, he narrates that upon employing this method, one native cow is capable of catering to the need of about thirty acres of land. It is also aimed to achieve the attainment of soil health restoration, promoting soil organic carbon while curbing the application of farmyard manure as is largely done in organic agriculture.

Economic Survey (2019) emphasized ZBNF as one of the threshold practices for improving income in the light of declining reagents and farm incomes. Almost symbiotic since great attention is devoted to microbial aspects (Gilbert et al., 2012) and "probiotic" functions in nature (Lorimer, 2017), they are other essential biological sciences governing soil and plant life stimulating the ecological renaissance in agriculture. Wallenstein (2017) staunchly argues in favor of nourishing soil microbes for rejuvenating soils. This, he explains, can be achieved by building up organic matter in the soil, reducing tillage, and abolishing the use of synthetic fertilizers and chemicals. Natural farming was first posited by Japanese farmer Masanobu Fukuoka, based on the philosophy of working in harmony with natural cycles and processes of the natural world (Fukuoka, 1987). It is viewed as the ultimate answer to end reliance on purchased inputs, enhance family health & nutrition, and provide stability in employment. This should convert into higher crop yield and thus reduced indebtedness and suicides among Indian farmers.

More than 85% of the total 146.5 million farmers of India are small holders, and more than 100 million farmers, or 68.5% of total farmers, farm an average of 0.38 ha of land in India. Hazell and Rahman reiterated that most of the poor and hungry people in the world live and work on small farms and scarcely survive on little land using low-input-low-yield technologies. In this scenario, the use of modern technology and innovation in Indian agriculture is being viewed as the only way out. Further, a segment of the scientific community and critics has vehemently opposed this alternative practice, condemning it for being not based on scientific evidence and for promoting certain belief systems, particularly concerning indigenous cows and a backward-looking and perhaps chauvinistic idiom.

In India, the National Academy of Agricultural Sciences (NAAS) held a one-day brainstorming session at Delhi and suggested that the government should not spend monetary and human resources in promoting ZBNF. The practice has also been condemned by calling it "an unproven" technology that shall not bring intangible gain to quite either the farmer or the consumer. The Indian Council of Agricultural Research (ICAR) has formed a committee with Prof. V. Praveen Rao, Vice Chancellor, PJTSAU, Hyderabad, as chairperson, to put ZBNF viability to review. The committee is conducting experimentation at five different locations in India. On the contrary, the proponents of ZBNF say that conventional farming or chemical farming is degrading the land as all kinds of chemicals are sprayed on the soil and into food systems, while agro-ecological systems replenish soil fertility. Munster strongly believes that after doing extensive research on the movement, the current ambivalence renders Natural Farming an excellent case for the political ecology of agriculture. This has since been reaffirmed by many Indian economists, among them the Finance Minister, who had stressed ZBNF policies and urged farmers to replicate this innovative model.

While addressing the 14th Conference of Parties (COP) to the UN Convention to Combat Desertification, the Prime Minister of India noted ZBNF as a way toward sustainable agriculture. Interestingly, the Government of Andhra Pradesh initiated a new project to improve the livelihoods of farmers with CRZBNF, which was later modified to Andhra Pradesh ZBNF in 2015 to combat climate change in the drought-prone Rayalaseema region (Ananthapur, Prakasam, Kadapa, Kurnool, and Chittoor). In pilot phases, 50 villages from 13 districts of the state had been selected but later were extended to cover the entire state (Niyogi, 2018). Besides, the state had adopted Community Managed Sustainable Agriculture (CMSA), wherein the use of chemical pesticides was replaced by a combination of physical and biological measures that incorporated bio-pesticides. This initiative aimed to reduce the use of chemical fertilizers based on chemical chemistry. CMSA was adopted by over 3, 00, 000 farmers in the state of Andhra Pradesh, covering about 1.36 million acres of farmland (Kumar et al., 2009). This same functional structure of CMSA has since been adopted to promote ZBNF practices in the state. As of 31/12/2019, the Andhra Pradesh ZBNF reported that 5.80 lakh farmers have implemented ZBNF practices on 2.60 lakh ha of cultivated area in 3011 villages across Andhra Pradesh (apzbnf.in).

There are various types of Natural Farming in which farmers do local adaptations and use locally available resources to enhance that particular practice, and hence, the stepping-stone for ZBNF or Natural Farming in India is from the advocacy of its chief proponent, Sh.Subhash Palekar. He has stirred several controversies in due course. Initially, he named the practice "Zero Budget Natural Farming" (ZBNF), which was later changed to "Zero Budget Spiritual Farming" (ZBSF), and sometimes, again renamed "Subhas Palekar Natural Farming" (SPNF). Worldwide, soils contain more carbon than the combined amounts in plants and in the atmosphere. When soils lose their carbon-rich organic matter, they release carbon dioxide, a greenhouse gas that may accelerate warming of the climate. However, through soil regeneration, one could sequester more carbon deep down in the ground, thus slowing down the global warming. Besides preserving soil, cover crops remove carbon from the atmosphere while they are growing and return it to the soil. In contrast to cash crops that are harvested and taken off from the soil, cover crops are allowed to decompose and develop soil. Plants are the original source of soil carbon; microbes determine its fate in the soil as food, thereby ensuring that at least part of it remains in the soil (Wallenstein, 2017). Thus, it is believed that ZBNF or Natural Farming rests on the above basis. With the introduction of some minor tweaks under it- application of microbes, use of cover crops, minimum tillage, multi-cropping, et cetera-it leads to the regeneration of soil and ultimately toward sustainable agricultural development.

While this market for organic food and beverages is continually growing, the story of organic farming is a controversial one, as well inefficient in production. Even though the increasing trend of demand among consumers had severely limited availability of organic products recently was largely due to the inability of supply to meet the rapidly increasing demand for organic foods (Dimitri and Oberholtzer , 2009). According to the World of Organic Agriculture 2018 report, 36% of the organic producers worldwide are in India, but India contributes only 2.59% (1.6 million hectares) of the total area of organic cultivation, which is approximately 57.8 million hectares across the globe. Because organic products require greater labor, certification, handling costs, and also there are relatively lower yields concerned, generally this means organic products are 3-4 times more costly. A farmer wishing to shift to organic farming must undergo a three-year transition whereby he practices organic farming but cannot sell his produce as organic. The transition is considered purely a financial risk for farmers in that they are reducing yields. If NF becomes chemically untainted production systems in law, for every sale during the transitional period, farmer will sell it as ‘Green Product’ straightaway from the first year itself at a little endowed price. This would help to even out compensation for the loss of yield, if any at all, during the initial years. According to a report by Tech Sci Research, the global organic food market was 110.25 billion dollars in 2016 and forecasted to grow at a CAGR of 16.15%, reaching 262.85 billion dollars by 2022. The market for organic foods in India has been growing at a CAGR of 25%.

India exported organic products worth □5,151 crores in 2018-19, around 50% more than the previous fiscal. Presently 1.78 million hectares under organic farming in India. But the exorbitant prices of organic food products in comparison with conventional foods make them out of reach for middle-class people. These high prices are a result of high input costs, labor, and separate handling charges along with the cost of certification, and the low yield during the conversion period of three years during which farmers must do organic farming but cannot sell the produce as organic. The high price of organic products is also deemed owing to the logistics involved in raising these products from certified organic farms within the city, which raises product costs. Yet another thought states that organic agriculture can only carry a population of 3-4 billion at maximum (Connor, 2008), much less than the present World population of 6.2 billion, much less than an anticipated 2050 figure of 9 billion. Badgley et al. (2007) show that organic farming will not only enhance crop productivity in developing countries but will, in fact, have the potential to feed the whole world. Barbieri et al. (2019) calculated the potential crop substitution based on organic farming.

31% decrease in harvested area was reported due to an increase in temporary fodder harvested areas (+63%), secondary cereals +27% and pulses +26% after compensating with major cereals like wheat, rice and maize, compared to regular conditions. These changes in combination with higher yields of organic crops compared to conventional ones may result in a −27% loss in energy production from croplands against current production. Searchinger et al. (2018) also contend that organically farmed food has a more severe carbon impact than conventionally farmed food because of the larger land areas that would be required. This is mainly because much lower crop yield is harvested, mainly for the reason that fertilisers are not used. We need a much larger area of land in order to produce the same amount of organic food, and thus this would indirectly raise carbon emissions. However, with regards to the point of view of consumers, organic food becomes more climate friendly. Contrary to these rational arguments brought forward in favour of organic farming, Sh. Subhash Palekar contests against organic farming as an alternative to conventional, posing more peril to Indian agriculture.

Organic farming needs a larger quantity of organic matter, that is, 'FYM' which most farmers in India lack. Therefore they have to buy cow dung at a larger scale, adding to the cost and making agriculture uneconomical. Hence organic produce has turned into a luxury product only for the rich, who can afford to buy it at exorbitant prices. There are many governmental and NGOs that are promoting vermi-compost for organic farming though by using a species of earthworm called Eisenia foetida, the redworm. However, it being an earthworm genus, it is a surfacefeeder and feeds only on the organic matter on the surface soil, thus degrading dried vegetation, compost, or manure. They do not work their way into the soil as local earthworms do and hence cannot provide the deep soil with castings that are the richest stock of minerals required by plants. Therefore, in the Indian context, the natural farming practice is more useful than organic farming or conventional input-intensive farming.

2

Principles of Natural Farming

1. No Tillage (No Ploughing or Turning of the Soil)

- The principle of No Tillage in natural farming refers to the practice of avoiding the mechanical disturbance of the soil, such as ploughing, turning, or tilling. This approach is central to natural farming because it allows the soil to maintain its natural structure, preserves its biodiversity, and enhances its ability to support plant growth sustainably.
- Tillage disrupts the natural layers of the soil. When the soil is left undisturbed, its structure remains intact, allowing for better water infiltration, root penetration, and nutrient availability. A well-structured soil is more resilient to erosion and compaction.
- Soil is home to a vast network of microorganisms, including bacteria, fungi, and earthworms, that play crucial roles in nutrient cycling and plant health. Tilling can destroy these delicate ecosystems, reducing the soil's fertility. No tillage preserves these beneficial organisms, promoting natural soil regeneration.
- In natural farming, organic matter (like dead plants, mulch, and crop residues) is left on the surface of the soil. As it decomposes, it contributes to the formation of humus, a rich organic material that improves soil fertility and helps retain moisture. Tillage disrupts this process, speeding up the decomposition of organic matter and reducing its long-term benefits.
- No tillage improves the soil's ability to absorb and retain water, which reduces the need for irrigation and helps crops withstand drought conditions. Additionally, undisturbed soil is more resistant to erosion because plant roots and soil structure act as a barrier against wind and water erosion.
- Tillage can release carbon stored in the soil into the atmosphere as CO_2, contributing to greenhouse gas emissions. No-till farming helps keep carbon sequestered in the soil, aiding in climate change mitigation.

- In conventional farming, tillage is often used to control weeds. In natural farming, however, weeds are managed through other means, such as mulching, cover cropping, and biodiversity. By leaving the soil undisturbed, weed seeds buried deep in the soil are less likely to germinate and grow.
- Plants grown in undisturbed soil develop deeper, stronger root systems, which makes them more resilient to environmental stresses such as drought and pests. They can access nutrients from deeper soil layers, improving their health and productivity.
- No-till farming reduces the need for machinery and fuel for plowing, making the farming process more energy-efficient. It also saves time and labor, as farmers don't need to engage in repeated tilling operations.

2. No Chemical Fertilizers

- The principle of No Chemical Fertilizers in natural farming is based on the idea that healthy, fertile soil should be naturally maintained without synthetic inputs. Instead of using chemical fertilizers, natural farming emphasizes restoring and nurturing the soil's organic matter, which helps sustain long-term productivity and environmental health.
- Natural farming relies on organic materials like compost, farmyard manure, crop residues, and green manure to enrich the soil. These organic inputs break down slowly, releasing nutrients in a form that plants can readily absorb. Organic matter also enhances soil structure and increases water retention, benefiting the entire ecosystem.
- Mulching involves covering the soil with plant residues (such as straw, leaves, or crop waste). This not only protects the soil from erosion and helps retain moisture but also provides a slow release of nutrients as the mulch decomposes, creating a natural nutrient cycle.
- Nitrogen is a critical nutrient for plant growth, and in natural farming, it is supplied through the cultivation of nitrogen-fixing plants like legumes. These plants form a symbiotic relationship with nitrogen-fixing bacteria in their roots, enriching the soil naturally without the need for synthetic nitrogen fertilizers.
- Cover crops, such as clover or rye, are grown during off-seasons to protect and nourish the soil. When these crops decompose, they add valuable nutrients and organic matter, enhancing soil fertility for future crops. This method also helps suppress weeds and prevent soil erosion.
- A healthy soil ecosystem is rich in microorganisms, fungi, and earthworms, which play a vital role in breaking down organic matter and

cycling nutrients. Natural farming practices foster the growth of these beneficial organisms by avoiding chemicals and maintaining organic matter in the soil.

- Natural farming encourages closed-loop systems where nutrients are recycled within the farm itself. Animal manure, plant residues, and other organic waste are returned to the soil, reducing the need for external inputs. This mimics the natural cycles found in forests and other ecosystems.

3. No Synthetic Pesticides or Herbicides

- The principle of No Synthetic Pesticides or Herbicides in natural farming is based on creating a balanced, self-sustaining ecosystem where pests and weeds are managed naturally. Instead of relying on synthetic chemicals, which can harm the environment and human health, natural farming employs biological and ecological methods to maintain crop health and manage unwanted plants and pests.
- Natural farming emphasizes biodiversity, both in terms of crops and surrounding ecosystems. A diverse farm environment creates habitats for beneficial insects and predators, which help control pest populations naturally. For example, ladybugs eat aphids, and birds or frogs might prey on insects that would otherwise harm crops.
- In a balanced ecosystem, pests have natural predators. Instead of killing pests with synthetic pesticides, natural farming encourages beneficial insects, spiders, birds, and animals that naturally reduce pest populations. This can be done by planting nectar-rich plants that attract these predators or creating habitats (like birdhouses or ponds) that encourage their presence.
- Companion planting involves growing plants together that benefit each other. Certain plants can repel pests or attract beneficial insects when grown alongside others. For instance, planting marigolds can help repel nematodes, while basil can deter flies and mosquitoes.
- Crop rotation helps break the life cycles of pests and diseases. By rotating crops seasonally or annually, pests that target specific crops are less likely to thrive, since their preferred hosts are no longer present in the same location.
- Natural farming sometimes employs biological pest control by introducing or encouraging species that feed on pests. For example, the introduction of parasitic wasps can help control caterpillar infestations, while nematodes can control soil-dwelling pests like grubs.

- Weeds are managed through methods like mulching and cover cropping rather than chemical herbicides. Mulch (such as straw, leaves, or wood chips) covers the soil, preventing light from reaching weed seeds and inhibiting their growth.
- In small-scale natural farming, manual weeding is often employed. While this requires labor, it avoids the need for chemical herbicides and maintains the integrity of the ecosystem.
- Some natural farming systems use plant-based, non-toxic pest control methods like neem oil, garlic sprays, or chili-based solutions. These are biodegradable and much safer for the environment, animals, and humans than synthetic pesticides.
- Healthy plants are more resistant to pests and diseases. Natural farming focuses on maintaining soil health through organic matter, composting, and mulching. Strong, resilient plants grown in nutrient-rich soil are less likely to succumb to pests or diseases, reducing the need for external pest management.
- Just as healthy soil supports plant growth, a rich microbial ecosystem can help combat pests and diseases. Natural farming promotes the growth of beneficial microorganisms that can outcompete harmful pathogens and protect plants from diseases.

4. No Weeding by Tillage or Herbicides

- The principle of **No Weeding by Tillage or Herbicides** in natural farming emphasizes the use of natural, non-invasive methods to manage weeds, rather than relying on tilling the soil or applying synthetic herbicides. Weeds are considered part of the ecosystem and are managed in ways that minimize harm to the soil, biodiversity, and surrounding environment. This approach encourages a more holistic management of weeds while maintaining soil health and promoting sustainability.
- Mulching involves covering the soil with organic or biodegradable materials, such as straw, leaves, grass clippings, or crop residues. Mulch prevents sunlight from reaching the soil, which inhibits weed seed germination and growth.
- Cover crops, such as legumes, grasses, or certain fast-growing plants, are sown to cover the soil during off-seasons or between planting cycles. These crops outcompete weeds by occupying space, using water and nutrients, and blocking sunlight.
- In natural farming, rotating crops and growing multiple types of plants together (polyculture) helps reduce weed pressure. Different plants have

varying root structures, growth habits, and nutrient needs, which prevent weeds from establishing dominance.

- Weeds are often less of an issue in ecosystems with high biodiversity. Natural farming fosters biodiversity by encouraging a variety of plant species, both in the main crops and in the surrounding environment (such as hedgerows or wildflower strips).
- In small-scale or intensive natural farming, manual weeding may still be practiced. Farmers use hand tools to remove weeds without disturbing the soil structure. While labor-intensive, manual weeding avoids the need for harmful chemical herbicides or machinery that disturbs the soil.
- Weeds are not always seen as enemies in natural farming. Some weeds provide benefits such as protecting the soil from erosion, attracting beneficial insects, or serving as indicators of soil conditions. For example, deep-rooted weeds can bring nutrients from deeper soil layers to the surface.
- Healthy, nutrient-rich soil can suppress weed growth naturally. Weeds often thrive in disturbed or nutrient-deficient soils. In natural farming, maintaining soil health through organic matter, composting, and natural mulches reduces the opportunities for weeds to establish and spread.
- Living mulch involves planting low-growing cover crops or ground cover plants between rows of main crops. These living mulches crowd out weeds and protect the soil, while still allowing space for the main crops to grow.
- Tilling disturbs the natural soil ecosystem, bringing buried weed seeds to the surface where they can germinate. By avoiding tillage, natural farming keeps weed seeds deeper in the soil, where they are less likely to sprout.

5. **Use of Mulching and Ground Cover**

- The Use of Mulching and Ground Cover is a foundational practice in natural farming that plays a crucial role in maintaining soil health, managing weeds, conserving moisture, and improving crop productivity. This practice is key to mimicking natural processes, where organic materials cover the ground, decompose, and nourish the soil. Mulching and ground cover support the no-tillage approach and help create a self-sustaining, balanced farm ecosystem.
- Mulching involves placing a layer of organic or biodegradable material on the soil surface around plants. The materials used can include straw, leaves, grass clippings, wood chips, rice husks, or crop residues.

Mulching is a versatile and essential technique that has several key benefits in natural farming.

- Mulch blocks sunlight from reaching the soil, preventing weed seeds from germinating and growing. This helps reduce the need for chemical herbicides or manual weeding.
- Mulch helps retain soil moisture by reducing evaporation. This is particularly important in dry climates or during periods of low rainfall. By conserving water, mulching reduces the need for frequent irrigation.
- Mulch insulates the soil, keeping it cooler in the summer and warmer in the winter. This provides a more stable environment for root growth and protects plants from extreme temperature fluctuations.
- Mulching protects the soil from erosion caused by rain, wind, or irrigation. The layer of mulch acts as a barrier, preventing soil particles from being washed or blown away.
- As organic mulch breaks down, it adds nutrients to the soil and improves soil structure. The decomposition process releases organic matter, fostering microbial activity and creating rich, healthy soil. Mulch also improves the soil's ability to retain nutrients and water.
- Mulching creates a favorable environment for earthworms and beneficial soil microorganisms. These organisms contribute to nutrient cycling, aerate the soil, and enhance overall soil fertility.
- Natural farming emphasizes using organic and locally available materials for mulching. Common types of mulch used in natural farming include:
  a. **Straw or Hay**: Widely available and effective at suppressing weeds and retaining moisture.
  b. **Leaves**: Fallen leaves are readily available on most farms and can be used to cover the soil surface.
  c. **Grass Clippings**: A byproduct of mowing, grass clippings can serve as mulch, although they should be used in thin layers to avoid matting.
  d. **Wood Chips or Bark**: These decompose more slowly than other organic materials and are good for long-term soil coverage.
  e. **Crop Residues**: Residues from previous crops (such as rice husks, corn stalks, or bean vines) can be left on the soil as mulch, returning nutrients to the soil.
- In addition to traditional mulching, **living mulch** and **ground cover plants** are used in natural farming to cover the soil. These are low-growing plants that spread across the ground, providing many of the

same benefits as organic mulch but with the added advantage of being part of the farm's biodiversity.

- Living mulch and ground cover plants compete with weeds for light, nutrients, and water, effectively suppressing their growth without chemicals.
- Like organic mulch, living mulch protects the soil from erosion and water loss, preserving soil structure and fertility.
- Certain ground cover plants, particularly legumes, fix nitrogen in the soil, improving its fertility for future crops. This natural form of fertilization reduces the need for chemical fertilizers.
- Ground cover plants contribute to on-farm biodiversity by attracting beneficial insects, pollinators, and soil organisms. They create a more balanced ecosystem that helps control pests and supports plant health.
- The root systems of living mulch plants contribute to soil aeration, organic matter buildup, and water infiltration, enhancing overall soil quality.
- Cover crops are a type of ground cover used primarily to improve soil health during off-seasons when the main crops are not growing. These crops, such as clover, vetch, rye, or mustard, are planted specifically to cover and protect the soil.
- Many cover crops, especially legumes like clover or beans, fix atmospheric nitrogen and convert it into a form plants can use. This naturally enriches the soil without synthetic fertilizers.
- When cover crops are cut down and left to decompose, they act as green manure, adding organic matter and nutrients to the soil, which enhances fertility and structure.
- Cover crops form dense growth that prevents weeds from gaining a foothold. This is particularly effective when cover crops are used in rotation with main crops.
- By covering bare soil during off-seasons, cover crops prevent erosion and protect soil from harsh weather conditions.
- Cover crops often have deep root systems that help break up compacted soil, improve water infiltration, and increase soil porosity.
- Mulching and ground cover plants contribute to the continuous cycling of nutrients in the soil. As organic matter decomposes or cover crops are turned into the soil, the nutrients are made available to future crops. This natural process maintains soil fertility without the need for external inputs like synthetic fertilizers.

- Mulching and living mulch systems can also help control pests and diseases. Certain mulch materials, like neem leaves, can have pest-repellent properties. Additionally, by covering the soil and reducing plant stress, mulching strengthens plant resistance to diseases and pests.

6. Diversity and Crop Rotation

- Diversity and Crop Rotation are key principles in natural farming, promoting ecological balance, enhancing soil health, and reducing pest and disease problems. These techniques imitate natural ecosystems, where multiple species coexist and enrich one another, contributing to the overall sustainability and productivity of the farm. Instead of monocultures, which can degrade soil and increase vulnerability to pests and diseases, natural farming uses diversity and crop rotation to create a resilient, self-sustaining system.
- In natural farming, **diversity** refers to cultivating multiple types of crops together rather than growing a single crop over large areas (monoculture). This practice is also known as **polyculture**.
- By planting a variety of crops, farmers create a more balanced and resilient ecosystem. Different plants interact with one another in ways that promote soil health, improve resource use, and control pests.
- Diverse crop systems reduce the spread of pests and diseases. In monocultures, pests or diseases that target a single crop can spread rapidly, but in a diverse system, these threats are less likely to find a suitable host, reducing the overall risk of infestation.
- Certain plants repel pests or attract beneficial insects. For example, planting marigolds alongside vegetables can help repel nematodes, while flowers like sunflowers attract pollinators that improve crop yields.
- Different plants have different nutrient needs and root structures. By planting a variety of crops, farmers ensure that no single nutrient is depleted from the soil. For example, deep-rooted plants like legumes can fix nitrogen and improve soil fertility, while shallow-rooted plants can access surface nutrients.
- Crop diversity encourages a healthy soil microbiome by fostering a wide range of beneficial soil organisms, including bacteria, fungi, and earthworms.
- In a diverse cropping system, plants with different growth habits and nutrient requirements share resources more efficiently. For example, tall plants can provide shade for shade-loving crops, while fast-growing

plants can cover the ground and reduce evaporation, conserving soil moisture.

- Mixed cropping systems also optimize the use of space and sunlight, allowing farmers to grow more food in the same area.
- By growing a variety of crops, farmers are less vulnerable to market fluctuations or crop failures. If one crop fails due to disease or adverse weather, other crops may still thrive, providing food and income security.
- **Crop rotation** involves planting different types of crops in a specific sequence over several seasons or years on the same piece of land. This practice prevents the depletion of soil nutrients, disrupts pest and disease cycles, and improves soil structure.
- Certain crops, particularly legumes like peas and beans, fix nitrogen in the soil. When rotated with crops that have higher nitrogen needs (such as corn or wheat), legumes enrich the soil with nitrogen, reducing the need for synthetic fertilizers.
- Crop rotation also helps balance nutrient use. Different crops draw varying amounts of nutrients from the soil. For example, heavy feeders like corn can be followed by light feeders like lettuce, which prevents over-extraction of specific nutrients.
- Crop rotation breaks the life cycle of pests and diseases that tend to build up when the same crop is grown in the same place repeatedly. Pests that thrive on one crop will find it difficult to survive when a different crop is planted the following season.
- For example, rotating brassicas (like cabbage) with non-brassicas (such as legumes) reduces the buildup of pests like cabbage worms or clubroot disease, which specifically target brassicas.
- Crop rotation improves soil structure by alternating crops with different root systems. Deep-rooted crops, such as alfalfa or sunflowers, break up compacted soil layers, while shallow-rooted crops help stabilize the topsoil and prevent erosion.
- Certain crops (like grasses or cereals) add organic matter and improve soil tilth, while others (like legumes) enrich the soil with nitrogen.
- Rotating crops with different growth habits and planting schedules disrupts weed growth patterns. For example, cover crops or fast-growing plants can outcompete weeds for space, light, and nutrients, reducing weed pressure in subsequent seasons.
- Some crop rotations include smother crops, which are dense-growing plants that inhibit weed germination and growth.

- Rotating crops promotes biodiversity both above and below the ground. By changing the crop environment regularly, the soil microbiome becomes more diverse, fostering a range of beneficial microorganisms that improve soil health.
- Above ground, rotation encourages beneficial insect populations, such as pollinators and predators that help control pests.
- **Legume-Grain Rotation**: Legumes (like beans or lentils) fix nitrogen in the soil, which can then be used by nitrogen-hungry grains (like wheat or corn) in the following season.
- **Root Leaf Rotation**: Rotating deep-rooted crops (like carrots) with shallow-rooted crops (like lettuce) helps maintain soil structure and improves nutrient uptake.
- **Three-Year Rotation**: A common rotation might include a year of legumes, followed by grains, and then a fallow or cover crop year. This cycle improves soil fertility, reduces pest pressure, and enhances biodiversity.
- **Companion Planting** is a form of crop diversity where certain crops are planted together because they benefit each other. For example:
- **Corn, Beans, and Squash** (known as the "Three Sisters"): Corn provides support for climbing beans, beans fix nitrogen in the soil, and squash spreads along the ground, acting as a living mulch to suppress weeds and retain soil moisture.
- **Tomatoes and Basil**: Basil helps repel pests that might target tomatoes and can enhance the flavor of the tomatoes as well.
- Another approach to increasing diversity in natural farming is **agroforestry**, which integrates trees or shrubs with crops and livestock. Trees provide multiple benefits such as shade, windbreaks, and habitat for birds and beneficial insects. Trees also improve soil fertility by recycling nutrients from deep in the soil through their roots and leaves.

7. Natural Seed Selection

- Natural Seed Selection in natural farming is a crucial practice that focuses on selecting seeds that are adapted to local conditions, resilient, and capable of thriving without synthetic inputs like fertilizers or pesticides. Unlike conventional agriculture, which often relies on commercially produced hybrid or genetically modified seeds, natural farming promotes the use of locally adapted, open-pollinated, and heirloom seeds that are saved and selected from the previous harvest. This approach enhances the genetic diversity, sustainability, and self-reliance of farming systems.

- Seeds are selected from plants that have shown resilience and strong growth in local environmental conditions, such as climate, soil type, and rainfall patterns. This ensures that the plants from these seeds are well-suited to the farm's specific ecosystem.
- By choosing seeds from plants that have naturally adapted to local pests, diseases, and climatic stresses (e.g., drought or heat), farmers reduce the need for chemical inputs and irrigation. Over time, these seeds become more robust and better suited to the local environment.
- Natural farming encourages farmers to save seeds from their best-performing plants at the end of each growing season. Seed saving allows farmers to develop their own seed stock that is better adapted to their specific farm conditions.
- This practice also promotes self-sufficiency, as farmers are less reliant on external seed suppliers and can maintain control over their seed resources.
- Seed saving typically focuses on open-pollinated or heirloom varieties, as these maintain genetic consistency from one generation to the next, unlike hybrid seeds that may not produce uniform offspring.
- Natural farming favors the use of **heirloom** and **open-pollinated** seeds over hybrid or genetically modified varieties. Heirloom seeds are traditional varieties that have been passed down through generations and are valued for their flavor, nutritional quality, and adaptability.
- Open-pollinated seeds reproduce true-to-type, meaning that plants grown from these seeds will closely resemble their parent plants, preserving their traits for future generations. These seeds are genetically diverse, which enhances the resilience of the crop population.
- The selection process focuses on identifying the healthiest, most vigorous, and productive plants in a crop. These plants are chosen based on several criteria, such as:
    - **Disease resistance**: Plants that remain healthy and productive even in the presence of pests and diseases are preferred.
    - **Yield**: High-yielding plants that produce a large quantity of fruits, grains, or vegetables are favored.
    - **Drought or flood tolerance**: In regions prone to extreme weather conditions, plants that perform well in drought or flooding conditions are prioritized.

  - **Nutritional value and flavor**: Plants that produce highly nutritious and flavorful fruits or vegetables are selected to ensure high-quality produce for consumption.
- A key goal of natural seed selection is to maintain and enhance genetic diversity within crop populations. Genetic diversity makes crops more resilient to pests, diseases, and environmental changes, reducing the risk of total crop failure.
- By selecting and saving seeds from a wide variety of plants, farmers preserve different traits, such as disease resistance, drought tolerance, and yield, which may be useful in future growing seasons. This contrasts with monocultures, where genetic uniformity can lead to vulnerability to a single pest or disease.
- Seeds selected through natural methods are adapted to low-input, sustainable farming practices. They are naturally suited to thrive without chemical fertilizers, pesticides, or herbicides.
- Over time, these seeds become accustomed to the ecological conditions of natural farming, including the presence of beneficial insects, microbes, and soil fertility practices like composting and mulching.
- In natural farming, the seed selection process is often a **farmer-led initiative**, where farmers themselves experiment, observe, and choose seeds that work best on their land. This participatory approach allows farmers to develop varieties that meet their specific needs, rather than relying on seeds developed in distant research centers.
- Through this process, farmers may create new landrace varieties—locally adapted populations of crops that develop over time through natural selection and human intervention.

- **Advantages of Natural Seed Selection**

Resilience to Local Conditions

By selecting seeds from plants that thrive in local conditions, farmers create crops that are better adapted to the specific climate, soil, and ecosystem. This leads to stronger, more resilient plants that require fewer external inputs like water or fertilizers.

Reduction in Dependency on External Seed Sources

Farmers who practice natural seed selection reduce their reliance on commercially available seeds, which are often expensive and may not be well-suited to local conditions. By saving and selecting their own seeds, farmers can become more self-reliant and save on seed costs.

Preservation of Biodiversity

Natural seed selection helps preserve and enhance biodiversity, both in the crops grown and in the surrounding ecosystem. By maintaining a diverse population of plants, farmers contribute to the conservation of genetic diversity, which is vital for food security and ecosystem health.

Cultural and Traditional Knowledge

Seed saving and selection are often rooted in cultural practices and traditional knowledge. Farmers pass down seeds and selection techniques through generations, preserving not only crop varieties but also valuable agricultural knowledge and heritage.

Customization for Farmer Needs

Through natural seed selection, farmers can tailor their crops to meet specific needs, such as resistance to local pests, improved flavor, or higher yields. This farmer-led approach empowers communities to develop crops that are well-suited to their particular needs and preferences.

Sustainability and Ecological Balance

Seeds that are selected and grown through natural farming methods are better suited to organic, low-input systems. These plants are more likely to thrive in an environment that prioritizes ecological balance, natural soil fertility, and minimal intervention.

Long-Term Crop Improvement

Over multiple generations, natural seed selection leads to continuous crop improvement. Each season, seeds are selected from the best-performing plants, gradually improving the overall resilience, yield, and quality of the crops.

- **Challenges of Natural Seed Selection**

Time-Intensive

Natural seed selection can be a slow process, requiring careful observation, record-keeping, and seed-saving techniques that take time to perfect.

Loss of Genetic Traits

Without careful management, there is a risk of losing valuable genetic traits if selection is not carried out systematically.

Cross-Pollination Risks

Open-pollinated plants may cross with other varieties, which can introduce unwanted traits or reduce the purity of the seed stock.

8. Water Conservation

- Water conservation is a critical component of natural farming, where the goal is to work in harmony with natural processes to optimize water use, reduce waste, and improve soil moisture retention. Instead of relying on external irrigation systems or over-extraction of water resources, natural farming employs techniques that enhance the soil's ability to capture, store, and use water efficiently. These practices not only conserve water but also promote healthy, resilient ecosystems that support long-term sustainability.
- Healthy soil with good structure plays a key role in water conservation. Soil rich in organic matter and microbial life has higher **water-holding capacity**, meaning it can absorb and retain water more effectively.
- **Composting** and the addition of organic matter, such as mulch, green manure, and cover crops, improve the soil's ability to hold moisture. Organic matter helps bind soil particles together, creating pore spaces that allow water to infiltrate and remain in the root zone for longer periods.
- **No-till farming** (another principle of natural farming) also improves water retention by minimizing soil disturbance and preserving soil structure. Undisturbed soil has better moisture retention, reduces evaporation, and allows plant roots to penetrate deeper.
- Mulching is a highly effective method for conserving soil moisture in natural farming. A thick layer of organic mulch (e.g., straw, leaves, grass clippings, or wood chips) is spread over the soil surface.
- **Cover crops** and ground cover plants (such as legumes or grasses) are used to protect the soil surface, reducing evaporation and improving water infiltration.
- These plants help break up compacted soil and improve soil structure, allowing water to penetrate deeper into the ground. Cover crops also protect the soil from erosion, particularly during heavy rains.
- Additionally, cover crops like clover or vetch contribute to nitrogen fixation, further enhancing soil fertility while conserving water.
- **Swales and contour bunding**: Swales are shallow ditches dug along the contour lines of a landscape to slow down and capture rainwater,

allowing it to infiltrate into the soil rather than run off. Contour bunds work similarly to slow down water and prevent soil erosion.

- **Ponds and reservoirs**: Farmers can dig ponds or reservoirs to collect and store excess rainwater, which can be used for irrigation during droughts.
- **Rainwater tanks**: In some systems, rainwater is collected from roofs and stored in tanks for later use in the fields or gardens.
- In natural farming, irrigation is used sparingly and only when absolutely necessary. When irrigation is needed, water-efficient systems like **drip irrigation** or **soaker hoses** are preferred because they deliver water directly to the plant's root zone, minimizing evaporation and runoff.
- These systems also allow farmers to apply small amounts of water over extended periods, promoting deep root growth and reducing water stress on plants.
- By focusing on deep-rooted crops and encouraging natural plant growth patterns, natural farming helps plants access water from deeper soil layers. Deep-rooted plants are less reliant on surface water and are better able to withstand drought conditions.
- Crops are chosen based on their ability to adapt to local conditions, and plants with strong, deep roots help to naturally aerate the soil and enhance its ability to hold water.
- Integrating trees into the farming system (agroforestry) plays a significant role in water conservation. Trees help improve water infiltration into the soil, reduce runoff, and prevent soil erosion.
- Tree roots also act as natural pumps, pulling water from deep within the ground and making it available to surrounding crops. In addition, trees provide shade, which reduces the temperature of the soil surface and limits evaporation.
- Trees and shrubs planted around the edges of fields can act as windbreaks, reducing the drying effects of wind on crops and soil.
- **Contour farming** involves planting crops along the natural contours of the land rather than in straight rows. This practice reduces soil erosion and water runoff by slowing down the movement of water across the landscape, allowing more time for water to infiltrate the soil.
- Contour bunds or terraces can also be built to trap water and prevent it from washing away topsoil. These structures help capture rainwater and direct it to where it is most needed in the field.

- In natural farming, farmers select crops that are naturally adapted to local conditions, including drought tolerance. By using varieties that require less water, farmers can reduce their dependence on irrigation and minimize water stress during dry periods.
- Many traditional and indigenous crop varieties have evolved to thrive with minimal water, making them ideal for natural farming systems in water-scarce regions.
- Natural farming promotes the creation of **microclimates** through diverse planting arrangements. Trees, shrubs, and other tall plants provide shade for lower-growing crops, reducing water evaporation from the soil and lowering plant stress during hot, dry weather.
- **Living fences** or windbreaks made of trees and shrubs can also be used to create sheltered environments that retain moisture and reduce water loss due to wind.

9. Livestock Integration

- Livestock integration in natural farming refers to the practice of incorporating animals into farming systems in a way that enhances both crop production and animal health while promoting ecological balance. This approach recognizes the symbiotic relationships between crops and livestock, utilizing the natural behaviors and contributions of animals to improve soil fertility, pest control, and overall farm sustainability.
- Livestock and crops can mutually benefit from each other when integrated effectively. Animals provide manure, which enriches the soil, while crops offer fodder and forage for the animals.
- This integration creates a closed-loop system where waste from one component becomes a resource for another, reducing the need for external inputs and enhancing the sustainability of the farm.
- Livestock play a crucial role in nutrient cycling. Their manure serves as a natural fertilizer, replenishing soil nutrients and enhancing soil health.
- By allowing animals to graze on crop residues, farmers can reduce waste while improving soil organic matter and nutrient content through manure deposition.
- Integrating livestock with crops allows animals to graze on a variety of plants, improving their diet and health. This practice can lead to better animal performance and higher-quality products.
- Multi-species grazing, where different types of animals graze together, can also enhance pasture health and reduce the risk of overgrazing on

any one species.

- Certain livestock species can act as natural pest control agents. For example, poultry can help control insect populations in fields, while pigs can root out pests in the soil.
- This reduces the need for synthetic pesticides and contributes to a healthier ecosystem.
- Livestock, especially those that graze or trample on the soil, can help aerate it, promoting better root growth and water infiltration.
- Their natural movements and behaviors can serve as a form of tillage, helping to mix organic matter into the soil and improve its structure without the need for mechanical tillage.
- Livestock can be part of an integrated pest management strategy. By grazing livestock in crop fields, farmers can reduce pest populations while providing the animals with fresh forage.
- This approach encourages the natural balance of pest and predator species in the ecosystem.
- Integrating livestock into cropping systems allows for multi-functional land use, maximizing the productivity of a given area.
- Farms can diversify their income streams by producing both crops and livestock products, leading to greater resilience against market fluctuations.

- **Methods of Livestock Integration**

Rotational Grazing

In rotational grazing, livestock are moved between different pastures or crop fields, allowing them to graze on diverse forage while giving other areas time to recover.This practice promotes pasture health, reduces overgrazing, and enhances soil fertility through manure distribution.

Silvopasture

Silvopasture involves integrating trees, forage, and livestock into a single system. Trees provide shade for livestock, improve biodiversity, and enhance soil moisture retention.The combination of trees and pasture can lead to higher overall productivity and increased resilience to climate variability.

Crop-Livestock Systems

In crop-livestock systems, crops and livestock are managed in a complementary manner. For example, animals can graze on crop residues after harvest, which helps manage waste and provide feed.Cover crops can also be grown to provide forage for livestock while enhancing soil health.

Integrative Feeding

Farmers can utilize by-products from crop production (such as leaves, stems, and other residues) as feed for livestock. This reduces waste and creates a more sustainable system.

Livestock can also be fed using pasture and forage crops, which reduces reliance on commercial feed.

Dairy and Poultry Integration

Integrating dairy cows or poultry into cropping systems can enhance nutrient cycling and improve soil fertility. Manure from these animals can be applied to fields, enriching the soil and supporting crop growth.For instance, using poultry litter as fertilizer can provide essential nutrients to crops while reducing waste.

• Advantages of Livestock Integration in Natural Farming

Enhanced Soil Fertility

The addition of livestock manure improves soil structure, fertility, and microbial activity, leading to healthier crops and better yields.

Reduced Chemical Inputs

By promoting natural pest control and nutrient cycling, integrated livestock systems reduce the need for synthetic fertilizers and pesticides, resulting in a more sustainable farming practice.

Improved Animal Health

Livestock that graze on diverse forages and have access to natural environments tend to be healthier and more productive, reducing veterinary costs and improving animal welfare.

Economic Resilience

Diversifying income streams through the integration of crops and livestock can make farms more resilient to market fluctuations and climate variability.

Biodiversity and Ecosystem Health

Integrating livestock promotes biodiversity in both crops and pasture systems, contributing to overall ecosystem health and resilience.

Sustainable Resource Use

Livestock integration promotes the efficient use of resources, creating a more sustainable system that conserves water and soil while producing food.

- **Challenges of Livestock Integration in Natural Farming**

Management Complexity

Managing both crops and livestock requires knowledge and skills in both areas, which can be challenging for some farmers.

Resource Competition

Livestock may compete with crops for resources, such as water and nutrients, if not managed properly. Careful planning is needed to ensure that both components benefit from the integration.

Infrastructure Needs

Farmers may need to invest in infrastructure, such as fencing, shelters, and watering systems, to effectively integrate livestock into their farming systems.

Disease Management

Introducing livestock into cropping systems can pose challenges in terms of disease management. Careful monitoring and biosecurity measures are essential to prevent the spread of diseases.

10. Focus on Soil Health

- Soil health is a fundamental principle of natural farming, which emphasizes the importance of nurturing and maintaining healthy soils as the foundation for sustainable agricultural practices. Healthy soil is critical for supporting plant growth, promoting biodiversity, and ensuring the long-term viability of farming systems. In natural farming, soil health is viewed holistically, considering not just physical properties but also biological and chemical aspects.

Soil Organic Matter

Building organic matter is essential for soil health. Organic matter improves soil structure, enhances water retention, and provides a habitat for beneficial microorganisms.

Practices such as composting, cover cropping, and adding green manures help increase soil organic matter levels, leading to improved soil fertility and resilience.

Diversity of Soil Life

Healthy soils are teeming with diverse life forms, including bacteria, fungi, earthworms, and other microorganisms. This biodiversity contributes to nutrient cycling, disease suppression, and overall soil health.Natural farming promotes practices that encourage soil life, such as reduced tillage and the use of organic amendments.

Minimal Disturbance

Minimizing soil disturbance is a cornerstone of natural farming. Practices like **no-till farming** help maintain soil structure and prevent erosion, allowing beneficial organisms to thrive.Disturbance from traditional tillage can disrupt soil aggregates, decrease organic matter, and harm soil microorganisms, leading to reduced soil health.

Cover Cropping

Growing cover crops is a critical practice in natural farming that helps protect and enhance soil health. Cover crops prevent soil erosion, improve soil structure, and increase organic matter content when they are terminated and incorporated into the soil.Additionally, cover crops help suppress weeds, fix nitrogen, and enhance microbial activity, contributing to overall soil fertility.

Crop Rotation and Diversity

Crop rotation and planting diverse crops promote soil health by disrupting pest and disease cycles, improving nutrient cycling, and enhancing soil structure. Different crops have varying root structures and nutrient requirements, which can help maintain balanced soil nutrients and reduce the risk of soil degradation.

Composting and Organic Amendments

Composting is an essential practice in natural farming that enriches soil with organic matter and nutrients. Compost adds beneficial microorganisms and improves soil structure, water retention, and nutrient availability.Other organic amendments, such as manure or biochar, also contribute to soil health by enhancing nutrient content and microbial diversity.

Soil Testing and Monitoring

Regular soil testing helps farmers understand soil nutrient levels, pH, and organic matter content, enabling them to make informed decisions about soil management practices. Monitoring soil health over time helps track improvements and identify any potential issues that need to be addressed.

Water Management

Effective water management practices support soil health by promoting proper drainage and preventing waterlogging or drought stress.Techniques such as mulching, contour farming, and the use of swales can help manage water efficiently, contributing to soil moisture retention and reducing erosion.

Integrating Livestock

Integrating livestock into farming systems can enhance soil health through natural fertilization and aeration. Manure from livestock adds organic matter and nutrients to the soil while their grazing can help control weeds. Careful management of livestock grazing prevents overgrazing and soil compaction, promoting healthier soils.

Natural Pest Management

Promoting soil health supports natural pest management by fostering biodiversity and encouraging beneficial insects that can help control pest populations. Healthy soils contribute to vigorous plant growth, making them more resilient to pests and diseases.

- **Benefits of Focusing on Soil Health**

Increased Crop Yields:

Healthy soils support robust plant growth, leading to higher crop yields and improved quality. Nutrient-rich, well-structured soils promote efficient nutrient uptake and root development.

Enhanced Resilience

Soils with good health are more resilient to environmental stresses, such as drought, flooding, and extreme temperatures. This resilience contributes to food security and sustainable production.

Reduced Input Costs

Healthy soils require fewer synthetic fertilizers and pesticides, leading to reduced input costs for farmers. This can result in greater profitability and sustainability.

Improved Water Retention

Soils rich in organic matter can hold more water, reducing the need for irrigation and helping to mitigate the effects of drought.

Biodiversity Preservation

Healthy soils promote diverse ecosystems, supporting a wide range of plants, animals, and beneficial microorganisms. This biodiversity enhances the resilience and productivity of the farm.

Climate Change Mitigation

Healthy soils can sequester carbon, helping to mitigate climate change by reducing greenhouse gas emissions. Practices that enhance soil health contribute to overall environmental sustainability.

• Challenges to Soil Health in Natural Farming

Soil Degradation:

Poor management practices, such as excessive tillage, monocropping, and chemical inputs, can lead to soil degradation, reducing soil health and productivity.

Knowledge and Skills

Farmers may require training and knowledge about soil health principles and practices, which can be a barrier to implementing natural farming methods.

Economic Constraints

Transitioning to natural farming practices focused on soil health may require initial investments in organic inputs, cover crops, or composting systems, which can be a barrier for some farmers.

3

Soil Health Management in Natural Farming

Natural farming is about keeping the long-term health of the soil in mind. Soil health is an active state or condition in which soil functions as a vital living ecosystem, enhancing plant, animal, and human health. Healthy soil has a good balance of minerals, organic matter, and moisture. With life processes occurring constantly, it has many complementary internal resources that allow it to be less reliant on external inputs. Natural farming helps to build this soil health, so that many years down the line, soil is in good health without being weakened by prolonged input redundancy. Regular tillage, which turns the soil over and breaks up the mycorrhizal fungi that connect the plants, is unnecessary in natural farming. We do not till. When we encourage life in the soil, it produces better soil heading towards fertility. Feeding the soil is the goal of natural farming. Ensuring robust health in the soil will, in turn, produce healthy plants, and by extension, a nutritious offering of produce. Soil health is a critical factor for natural farming. In every action we take to improve the health of the soil, we are addressing a component of healthy farming and being kind to the land. Soil health is the foundation of natural farming. The following sections detail the importance of organic matter, the role of microorganisms, water conservation, and preparation of natural inputs. By understanding these facts, we can truly understand soil health. Commercial inputs like pesticides, herbicides, and chemical fertilizers cost money. Natural farming aims to avoid the profit-zapping reliance on these costly inputs. Healthy soil produces hardy crops; strong plants do not get sick. Kept at their peak of health, strong, life-filled soils have a congress of life in their root zones, providing steady food and balance to crops. There are no external inputs constantly taxing the health of these crops. By tending to the farm as a living system, soils and plants are helped to be as healthy as possible, of their own accord.

1. Minimal Tillage or No-Till Farming

In natural farming, minimal tillage or no-till farming is a key practice aimed at preserving the natural structure and health of the soil. Unlike conventional farming, where the soil is extensively tilled to prepare the seedbed, natural farming focuses on maintaining soil integrity by reducing or eliminating tillage. This approach helps protect the soil's organic matter, prevents erosion, and supports beneficial microbial communities that are essential for nutrient cycling. In no-till systems, crop residues from previous harvests are left on the soil surface. These residues act as natural mulch, helping to retain moisture, suppress weeds, and gradually decompose, enriching the soil with organic matter and nutrients. The reduced soil disturbance also helps maintain earthworms and fungi, which improve soil aeration and structure, promoting healthier root systems. Minimal tillage or no-till farming aligns with natural farming's goal of working with nature to regenerate soil fertility. By avoiding heavy machinery and synthetic inputs, these practices promote biodiversity, enhance water retention, and reduce greenhouse gas emissions through carbon sequestration. Overall, minimal tillage fosters a self-sustaining, resilient agricultural system that supports long-term soil health, making it a vital component of natural farming's holistic approach.

2. Cover Cropping

In natural farming, cover cropping is a vital practice aimed at protecting and enriching the soil naturally. Cover crops, such as legumes, grasses, or cereals, are grown between main crop cycles or alongside primary crops to maintain a living plant cover on the soil. Unlike conventional farming, where fields are often left bare between growing seasons, natural farming uses cover crops to promote continuous soil health. Cover crops help prevent soil erosion by holding the soil in place with their root systems. They also enhance soil fertility by adding organic matter as they decompose. Leguminous cover crops, like clover or beans, have the added benefit of fixing nitrogen from the atmosphere, enriching the soil without the need for synthetic fertilizers. In addition to improving nutrient cycling, cover cropping in natural farming enhances water retention, prevents weed growth, and fosters beneficial soil microorganisms. These plants also provide habitats for pollinators and predatory insects that control pests, contributing to a balanced ecosystem. By maintaining biodiversity and building up soil organic matter, cover cropping aligns with the principles of natural farming—working with nature to create a sustainable, self-sufficient agricultural system that improves soil health and productivity over time without chemical inputs.

3. Composting and Organic Amendments

In natural farming, composting and organic amendments are essential for maintaining soil fertility and enhancing its biological activity without the use of synthetic fertilizers. Composting involves the controlled decomposition of organic materials like crop residues, kitchen waste, animal manure, and plant matter into nutrient-rich humus. This compost is then added to the soil as a natural fertilizer. Organic amendments such as compost provide a steady release of nutrients to plants, improve soil structure, and enhance moisture retention. They also promote the growth of beneficial microorganisms, fungi, and earthworms, which play crucial roles in breaking down organic matter and making nutrients more accessible to plants. By enriching the soil with organic matter, composting helps improve soil aeration and prevents compaction, allowing roots to penetrate more easily. This leads to healthier plant growth and greater resilience against environmental stressors like drought or heavy rain. In natural farming, composting and organic amendments are part of a closed-loop system where waste is recycled back into the soil. This reduces the need for external inputs, conserves resources, and aligns with natural farming's principles of sustainability and self-sufficiency. Overall, composting restores soil health, improves fertility, and supports long-term agricultural productivity.

4. Mulching

Mulching is a key practice in natural farming that involves covering the soil with organic materials like straw, leaves, grass clippings, or crop residues to protect and nourish the soil. In natural farming, mulching plays a vital role in mimicking the forest floor, where organic matter continually falls, decomposes, and enriches the soil. One of the primary benefits of mulching is moisture conservation. By covering the soil, mulch reduces evaporation, helping to maintain consistent soil moisture levels, which is particularly valuable in water-scarce regions. Mulch also regulates soil temperature, keeping it cooler in hot weather and warmer during colder periods, thus providing a more stable environment for plant roots. Mulching suppresses weed growth by blocking sunlight, which reduces the need for weeding or herbicides. As organic mulch decomposes, it slowly releases nutrients into the soil, enriching its fertility. This process also promotes the growth of beneficial soil organisms, such as earthworms and microbes, which are crucial for nutrient cycling and soil health. In natural farming, mulching not only protects the soil from erosion and nutrient loss but also contributes to building organic matter, improving soil structure, and creating a sustainable, self-renewing agricultural system that enhances productivity over time.

5. Crop Rotation and Polyculture

Crop rotation involves planting different crops in a planned sequence over time on the same land. This practice prevents nutrient depletion, as different crops have varying nutrient requirements and contributions. For instance, legumes, like clover or beans, can fix atmospheric nitrogen, enriching the soil for subsequent nitrogen-hungry crops like corn. By disrupting pest and disease cycles, crop rotation also reduces the risk of infestations and diseases that thrive on monocultured plants. Furthermore, varying root structures from different crops improve soil structure and aeration, enhancing water infiltration and reducing erosion.

Polyculture entails growing multiple crops simultaneously in the same area, promoting plant diversity and ecological balance. This practice enhances nutrient cycling, as different plants utilize and replenish soil nutrients in complementary ways. Polyculture can also suppress weeds, as diverse crops compete for resources, reducing the need for chemical herbicides. Additionally, a mix of plants attracts beneficial insects that help control pests naturally.

6. Use of Natural Fertilizers

In natural farming, the use of natural fertilizers is a fundamental practice that aligns with the philosophy of enhancing soil health and fertility without relying on synthetic chemicals. Natural fertilizers are derived from organic materials and are designed to enrich the soil and support plant growth through the addition of essential nutrients and the promotion of beneficial soil microorganisms.

Types of Natural Fertilizers

1. **Compost**: Decomposed organic matter, such as kitchen scraps, plant residues, and animal manure, compost provides a balanced source of nutrients and improves soil structure.
2. **Green Manure**: Cover crops, such as clover or alfalfa, are grown specifically to be tilled back into the soil. These crops add organic matter and can fix nitrogen, enhancing soil fertility.
3. **Animal Manure**: Well-composted manure from livestock is rich in nutrients and organic matter. It enhances microbial activity and improves soil structure and water retention.
4. **Bone Meal and Blood Meal**: These organic amendments provide concentrated sources of phosphorus and nitrogen, respectively, supporting strong root and foliage development.
5. **Fish Emulsion**: This liquid fertilizer made from fish waste is rich in nutrients and promotes healthy plant growth while improving soil microbial activity.

Benefits of Natural Fertilizers

- **Improved Soil Health**: Natural fertilizers enhance soil structure, promote microbial diversity, and increase organic matter, leading to healthier soils over time.
- **Nutrient Cycling**: They support the natural nutrient cycle, providing plants with essential nutrients in a form that is readily available.
- **Environmental Sustainability**: Natural fertilizers reduce the risk of chemical runoff and pollution, promoting a healthier ecosystem.
- **Reduced Dependency on Chemicals**: By using natural fertilizers, farmers can reduce their reliance on synthetic inputs, aligning with the principles of sustainability and self-sufficiency.

7. Agroforestry

Agroforestry is an integral practice in natural farming that combines agricultural and forestry practices to create sustainable, productive ecosystems. This method involves integrating trees, shrubs, and other perennial plants with crops and livestock in the same land management system. Agroforestry mimics natural ecosystems, promoting biodiversity, enhancing soil health, and providing various environmental and economic benefits.

Key Components of Agroforestry in Natural Farming

1. **Tree Cropping**: Planting fruit or nut trees alongside annual crops can yield multiple products from the same land, such as fruits, nuts, and grains. This diversifies income and provides shade and habitat for beneficial wildlife.
2. **Alley Cropping**: In this system, rows of trees or shrubs are planted alongside rows of crops. The trees provide shade, reduce wind erosion, and enhance soil moisture retention, while their leaves can be used as organic mulch.
3. **Silvopasture**: This practice combines trees, pasture, and livestock. Livestock graze under trees, benefiting from shade and forage, while trees contribute to soil health and provide additional income from timber or fruit.
4. **Forest Farming**: This involves cultivating high-value crops under the protection of a forest canopy, such as medicinal herbs or mushrooms, allowing farmers to diversify their production.

Benefits of Agroforestry

- **Biodiversity Enhancement**: Agroforestry systems support a variety of species, contributing to increased resilience against pests and diseases and promoting ecosystem health.
- **Soil Improvement**: Tree roots enhance soil structure and prevent erosion, while leaf litter adds organic matter, improving soil fertility and moisture retention.
- **Climate Resilience**: The diversity of plants and trees can help mitigate the impacts of climate change by improving carbon sequestration, reducing runoff, and providing shade during extreme weather.
- **Economic Diversification**: Farmers can generate multiple income streams from timber, fruits, nuts, and crops, making their operations more economically viable.

8. Water Management

Water management is a critical aspect of soil health management in natural farming, as it directly influences soil moisture, nutrient availability, and overall crop productivity. Effective water management strategies in natural farming focus on conserving water resources, enhancing soil moisture retention, and promoting sustainable practices that align with the principles of ecological balance.

Key Strategies for Water Management in Natural Farming

1. **Rainwater Harvesting**: Collecting and storing rainwater helps provide a reliable water source during dry periods. This can be done through the construction of ponds, swales, or rainwater catchment systems that capture runoff from roofs and hard surfaces.
2. **Contour Farming**: Implementing contour plowing and planting along the natural contours of the land reduces soil erosion and enhances water infiltration. This technique slows down water runoff, allowing more time for the soil to absorb moisture.
3. **Cover Cropping and Mulching**: Growing cover crops and applying organic mulch on the soil surface significantly improve moisture retention. Cover crops protect the soil from evaporation, while mulch acts as a barrier that reduces water loss and regulates soil temperature.
4. **Drip Irrigation**: Using drip irrigation systems delivers water directly to the root zone of plants, minimizing evaporation and runoff. This method promotes efficient water use and reduces the risk of over-saturation, which can lead to nutrient leaching.

5. **Soil Improvement Practices**: Enhancing soil structure through composting, minimal tillage, and organic amendments increases the soil's ability to retain moisture. Healthy soils with good organic matter content can hold more water and provide a stable environment for plants.
6. **Agroecological Practices**: Integrating diverse plant species through polyculture and agroforestry improves water management by creating microclimates that help regulate temperature and moisture levels. The roots of various plants can also enhance water infiltration and retention.

Benefits of Effective Water Management

- **Soil Health Improvement**: Proper water management enhances soil structure and microbial activity, promoting nutrient cycling and fertility.
- **Drought Resilience**: Sustainable water practices increase soil moisture availability during dry periods, helping crops withstand drought conditions.
- **Erosion Control**: By reducing runoff and promoting water infiltration, effective water management minimizes soil erosion and degradation.
- **Increased Productivity**: Consistent moisture levels support healthy plant growth, leading to higher yields and better-quality crops.

9. Green Manuring

Green manuring is a key practice in soil health management within natural farming that involves growing specific crops—known as green manure crops—with the intention of tilling them back into the soil before they mature. This practice enhances soil fertility, structure, and biodiversity while aligning with the principles of sustainable agriculture.

Key Aspects of Green Manuring in Natural Farming

1. **Types of Green Manure Crops**: Common green manure crops include legumes (such as clover, vetch, and alfalfa), grasses (like rye and buckwheat), and certain brassicas. Leguminous plants are particularly valuable as they have the ability to fix atmospheric nitrogen, enriching the soil with this essential nutrient.
2. **Nutrient Enrichment**: When green manure crops are incorporated into the soil, they decompose and release nutrients, especially nitrogen, phosphorus, and potassium. This natural fertilization reduces the need for synthetic fertilizers and enhances nutrient availability for subsequent crops.

3. **Soil Structure Improvement**: The root systems of green manure crops help improve soil structure by creating channels for air and water infiltration. This enhances soil aeration and reduces compaction, allowing roots of future crops to penetrate more easily.
4. **Organic Matter Addition**: Incorporating green manures adds organic matter to the soil, which improves moisture retention, promotes microbial activity, and enhances soil fertility over time. This organic matter is essential for building healthy, resilient soils.
5. **Weed Suppression**: Growing green manure crops can effectively suppress weeds by competing for light, space, and nutrients. This reduces the reliance on chemical herbicides, aligning with the principles of natural farming.
6. **Pest and Disease Management**: Certain green manure crops can help disrupt pest and disease cycles. For example, planting specific legumes can deter harmful nematodes and promote beneficial insects.

Benefits of Green Manuring

- **Sustainable Nutrient Cycling**: Green manuring promotes a closed-loop nutrient system, reducing the dependency on external inputs and enhancing soil health.
- **Enhanced Soil Biodiversity**: The addition of organic matter and nutrients fosters a diverse microbial community, which is crucial for nutrient cycling and plant health.
- **Improved Resilience**: Soils enriched with green manures are more resilient to environmental stressors such as drought, erosion, and nutrient depletion.
- **Cost-Effectiveness**: By reducing the need for synthetic fertilizers and herbicides, green manuring can lower overall production costs for farmers.

10. Biological Pest Control

Biological pest control is a fundamental aspect of soil health management in natural farming, emphasizing the use of natural predators, parasites, and pathogens to manage pest populations. This approach aligns with the principles of sustainability and ecological balance, reducing the reliance on synthetic pesticides and promoting a healthier agroecosystem.

Key Components of Biological Pest Control in Natural Farming

1. **Natural Predators**: Encouraging the presence of beneficial insects, such as ladybugs, lacewings, and predatory beetles, can help control pest populations. These natural predators feed on common agricultural pests like aphids, caterpillars, and mites, thereby reducing the need for chemical interventions.
2. **Parasitoids**: These are organisms that lay their eggs in or on pests, ultimately killing them. For example, parasitic wasps can target caterpillars or aphids, providing a natural means of pest control. Promoting habitats for these parasitoids, such as flowering plants, can enhance their populations.
3. **Microbial Agents**: Utilizing beneficial microorganisms, such as bacteria (e.g., *Bacillus thuringiensis* or Bt) and fungi (e.g., *Beauveria bassiana*), can effectively control pest populations. These microbes can infect and kill pests without harming beneficial insects or the environment.
4. **Companion Planting**: Certain plants can repel pests or attract beneficial insects. For instance, marigolds can deter nematodes, while plants like basil can attract pollinators and predatory insects. Strategic planting of companion crops helps enhance biodiversity and pest resistance.
5. **Crop Diversity**: Practicing crop rotation and polyculture increases biodiversity, which disrupts pest life cycles and reduces the likelihood of pest outbreaks. A diverse planting strategy creates a more resilient ecosystem, making it harder for pests to thrive.

Benefits of Biological Pest Control

- **Soil Health Improvement**: Biological pest control contributes to a balanced ecosystem, promoting the health of soil organisms that are essential for nutrient cycling and soil fertility.
- **Reduced Chemical Use**: By relying on natural pest control methods, farmers can minimize the use of synthetic pesticides, which can harm beneficial organisms and degrade soil health.
- **Resilience Against Pests**: Natural pest control methods help build a resilient agroecosystem that can better withstand pest outbreaks and environmental stressors.
- **Sustainability**: Biological pest control aligns with the principles of natural farming by promoting sustainable agricultural practices that enhance ecological balance and biodiversity.

4

Four Pillars of Natural Farming

A. Beejamruit

Beejamrit is an organic solution primarily used in seed treatment to promote better germination, protect against soil-borne diseases, and enhance plant growth. It is part of traditional agricultural practices in India and is commonly used in organic farming. he preparation of Beejamrit involves simple, locally available ingredients like cow dung, cow urine, lime, water, and a handful of soil (often sourced from under a banyan tree). The process is straightforward: cow dung is soaked in water overnight, and the mixture is later combined with lime, cow urine, and soil to create the final solution. This solution is then used to coat seeds before planting, providing them with a protective layer that boosts germination and shields them from pathogens. Beejamrit is valued for its low cost and quick preparation compared to chemical treatments. It is rich in beneficial microorganisms that help improve soil health and support plant growth.

The preparation of Beejamrit, a traditional organic solution used for seed treatment, is simple and involves locally available ingredients. Here's the process:

Ingredients

- Cow dung (local/indigenous): 5 kg
- Cow urine: 5 liters
- Lime: 50 grams
- Water: 20 liters
- Handful of soil (preferably from under a banyan tree)

Steps for Preparation

1. Soak cow dung: Place 5 kg of cow dung in a cloth, tie it up, and immerse it in 20 liters of water for about 12 hours.
2. Prepare lime water: Dissolve 50 grams of lime in 250 ml of water and leave it overnight.

3. Mix ingredients: After 12 hours, squeeze the cow dung into the water, releasing its nutrients.
4. Add other ingredients: Add 5 liters of cow urine, lime water, and a handful of soil to the cow dung water. Stir well using a wooden stick.
5. Ready for use: The solution is now Beejamrit and should be used within 24 hours.

Usage

- Seed Treatment: Soak the seeds or saplings in Beejamrit for a few minutes before planting, or coat the seeds by spraying the solution.
- Application: It can also be sprinkled around the roots of growing plants to prevent diseases.

The application of Beejamrit is primarily for seed treatment and plant root protection, and it can be used in multiple ways in organic farming:

1. Seed Treatment

- **Spraying:** After preparing Beejamrit, spread the seeds on a clean surface (cement floor or plastic sheet) and spray the solution evenly over them. Make sure all seeds are covered.
- **Dipping:** Seeds or saplings can be dipped in Beejamrit for a few minutes before planting. After dipping, let the seeds air dry in a shaded area before sowing.

2. Transplantation

- **Root Dip:** When transplanting saplings, dip their roots in Beejamrit for a few seconds. This helps protect them from fungal infections and encourages better root establishment

3. Soil Application

- **Root Zone Application:** Beejamrit can also be applied to the root zone of growing plants. This helps in enhancing soil microbial activity, which supports plant health and nutrient uptake

B. Jeevamruit

The soil contains nutrients, but they are inaccessible to plant roots. Jeevamrut, an organic fertilizer, contains beneficial microorganisms that convert these nutrients to a usable form when added to the soil. It can be sprayed or added to irrigation every 10-15 days until the soil improves. Jeevamrutha is an organic alternative to chemical fertilizers, rich in nutrients like carbon, nitrogen, phosphorus, and calcium. It boosts soil microorganisms, enhancing soil fertility and crop yield. It's made from cow dung and urine, promoting nutrient availability and microbial activity. This leads to more earthworms and fertile soil. Jeevamrutha is a bio-fertilizer that enhances plant growth by stimulating soil microbes and earthworms. It includes important microorganisms like rhizobacteria, cyanobacteria, mycorrhizal fungi, and nitrogen-fixing bacteria. This product catalyzes nutrient conversion and helps fight plant diseases. Subash Palekar recommends Jeevamrutha for nutrient conversion in plant roots. It contains PGPR, cyanobacteria, Solubilizing Bacteria (PSB), mycorrhizal fungi, and Nitrogenfixing bacteria, aiding nutrient absorption and disease control. It's effective for the first 3 years, after which the system stabilizes. Soil microorganisms actively impact fertility by cycling nutrients like carbon and nitrogen essential for plant growth. Jeevamrutha is a fermented culture that not only provides nutrients but also stimulates microbial activity and earthworms. It's beneficial against fungal and bacterial diseases. According to Palekar, it's necessary for the initial 3 years of transition to a self-sustaining system.

Ingredients

- **Cow dung** (from an indigenous cow): 10 kg
- **Cow urine**: 10 liters
- **Jaggery**: 1 kg
- **Pulse flour** (like gram flour): 1 kg
- **Handful of soil** (from the farm or termite mound)
- **Water**: 200 liters

Preparation Steps

1. **Mix ingredients**: In a large container, dissolve the cow dung, cow urine, jaggery, and pulse flour into 200 liters of water.
2. **Add soil**: Add a handful of soil from the farm (especially from termite mounds), which contains beneficial microbes.
3. **Fermentation**: Let the mixture ferment for about 3 to 5 days. Stir the mixture twice a day with a wooden stick to ensure proper aeration.
4. **Ready to use**: After fermentation, Jeevamrit is ready to be applied.

Application

- **Soil Fertility**: Jeevamrit can be applied to the soil every 7-15 days to improve soil health and provide plants with essential nutrients.
- **Foliar Spray**: It can also be used as a foliar spray to enhance plant growth and protect against diseases.

Benefits

- Increases microbial activity in the soil, making nutrients more available to plants.
- Improves plant growth, root development, and yields.
- Helps in the natural nitrogen fixation process and enhances soil health by increasing organic matter.

C. Mulching

Mulching is an agricultural and gardening practice that involves covering the soil surface around plants with a protective layer of material. It can be organic (natural) or inorganic (synthetic), and it serves multiple purposes.

Types of Mulch

1. Organic Mulch

- Composed of natural materials that decompose over time, adding nutrients to the soil.
- Examples: Straw, grass clippings, leaves, compost, bark, and wood chips.

2. Inorganic Mulch

- Made from non-biodegradable materials that do not decompose.
- Examples: Plastic sheets, landscape fabric, stones, and gravel.

Benefits of Mulching

1. **Moisture Retention**: Mulch reduces water evaporation from the soil, keeping it moist and reducing the need for frequent irrigation.
2. **Weed Suppression**: Mulch blocks sunlight, preventing weed seeds from germinating and growing.
3. **Soil Temperature Regulation**: It helps maintain consistent soil temperatures, keeping roots cooler in summer and warmer in winter.
4. **Soil Health Improvement**: Organic mulch breaks down over time, adding organic matter and nutrients to the soil.
5. **Erosion Control**: Mulching helps prevent soil erosion by reducing the impact of raindrops on the soil surface.
6. **Pest Control**: Some types of mulch, like pine needles, can act as natural repellents for certain pests.

Application

- Spread mulch evenly over the soil around plants, ensuring a 2-4 inch thick layer for organic mulch and a slightly thinner layer for inorganic mulch.
- Keep the mulch a few inches away from plant stems to prevent rot and pest infestation.

D. Whapasa

Whapasa is a concept in Zero Budget Natural Farming (ZBNF), an Indian sustainable farming approach pioneered by Subhash Palekar. It refers to maintaining an optimal balance between air (aeration) and moisture in the soil, which is crucial for healthy plant growth.

Key Principles of Whapasa

1. Balanced Moisture: Whapasa emphasizes maintaining soil moisture, but without waterlogging or completely saturating the soil. Instead of flooding the fields (as is common in traditional irrigation), the soil should retain moisture and have enough aeration for roots to breathe.
2. Aeration and Moisture: In the Whapasa technique, the soil must contain both air and water in appropriate proportions. This ensures that beneficial microorganisms in the soil are active and plants can access both nutrients and oxygen through their roots.
3. Reduced Irrigation: Unlike conventional methods that rely on frequent and heavy irrigation, Whapasa promotes light, controlled irrigation. This avoids excessive watering, reducing the need for large amounts of water while improving crop resilience.

Benefits of Whapasa

- Improved Soil Health: Proper moisture and aeration promote the growth of beneficial microorganisms and earthworms, improving soil structure and fertility.
- Reduced Water Usage: Whapasa reduces the need for irrigation by optimizing the soil's moisture-retaining capacity.
- Healthier Crops: Plants grown under Whapasa conditions have better root development and are more resistant to diseases and stress, leading to healthier, more robust crops.

5

Weed Management Under Natural Farming

Weed management in natural farming emphasizes natural, chemical-free methods that align with ecological principles. Instead of using herbicides or synthetic chemicals, natural farming promotes approaches that maintain biodiversity, improve soil health, and reduce weed pressure through various cultural, mechanical, and biological methods.

Here are key strategies for weed management in natural farming:

1. Mulching

Mulching is an agricultural and horticultural practice in which organic or inorganic materials are spread over the soil surface to protect and improve soil quality. In natural farming, mulching plays a key role by fostering a healthier ecosystem for plants without synthetic inputs.

Types of Mulch

Mulches are broadly categorized into **organic** and **inorganic** types:

- **Organic mulch** includes materials such as straw, leaves, grass clippings, crop residues, compost, wood chips, and sawdust. These materials decompose over time, enriching the soil with organic matter and nutrients.
- **Inorganic mulch** consists of materials like plastic sheeting, landscape fabric, and gravel. While these do not decompose, they effectively suppress weeds and retain moisture but do not improve soil fertility.

Benefits of Mulching

1. **Moisture Retention:** Mulch acts as a barrier that reduces water evaporation from the soil surface, helping to conserve moisture. This is particularly beneficial in drought-prone regions, as it allows farmers to reduce irrigation frequency and improve water-use efficiency.

2. **Weed Suppression:** Mulch prevents sunlight from reaching the soil, inhibiting weed seed germination. Organic mulches provide a double benefit, as the decomposing material can make it more difficult for weeds to grow, while also contributing to soil fertility.
3. **Soil Temperature Regulation:** Mulch acts as an insulating layer, keeping the soil cooler during hot weather and warmer during cold conditions. This temperature regulation creates a stable environment for root systems, enhancing plant growth and preventing stress from extreme temperatures.
4. **Soil Health Improvement:** Organic mulches decompose over time, adding organic matter to the soil. This enhances soil structure, promotes microbial activity, and improves nutrient cycling. As the mulch breaks down, it increases the soil's capacity to retain moisture and nutrients.
5. **Prevention of Soil Erosion:** Mulch provides a protective cover for the soil, reducing the impact of rain and wind that can cause erosion. This is especially important on sloped terrains where soil can easily wash away during heavy rains.
6. **Pest Control:** Certain types of mulch, like cedar or pine needles, can act as natural pest repellents. Mulching also creates a habitat for beneficial organisms such as earthworms and beetles that contribute to pest control and improve soil aeration.

2 Cover cropping is an agricultural practice in which specific crops, known as **cover crops**, are grown primarily for the benefit of the soil rather than for harvest. These crops are planted during fallow periods, between main crop cycles, or alongside main crops to protect and enhance the soil. Cover cropping is a vital component of natural and regenerative farming systems because it promotes soil health, improves biodiversity, and reduces the need for external inputs like chemical fertilizers or herbicides.

Types of Cover Crops

Cover crops can be classified into several types based on their primary functions:

1. **Legumes:** Crops like clover, alfalfa, and beans are nitrogen-fixing plants. They form a symbiotic relationship with soil bacteria to convert atmospheric nitrogen into a form that plants can use, enriching the soil with this essential nutrient.
2. **Grasses:** Rye, oats, barley, and sorghum are commonly used to prevent soil erosion and suppress weeds. Their extensive root systems hold soil in place and improve soil structure.

3. **Brassicas:** Crops like mustard, radishes, and turnips are known for their ability to break up compacted soils with their deep roots and suppress certain soil-borne diseases.
4. **Non-leguminous Broadleaves:** Examples include buckwheat and sunflower. These crops grow quickly and help smother weeds while providing organic matter to the soil.

Benefits of Cover Cropping

1. **Soil Fertility Improvement:** Cover crops, especially legumes, enhance soil fertility by fixing atmospheric nitrogen, reducing the need for synthetic fertilizers. When cover crops are plowed into the soil, they decompose, adding organic matter and nutrients, improving soil health for the subsequent crops.
2. **Erosion Control:** Cover crops provide a vegetative cover that protects the soil from erosion caused by wind and water. This is particularly beneficial on sloped or degraded land, where bare soil is more prone to being washed away during rains.
3. **Weed Suppression:** The dense growth of cover crops prevents sunlight from reaching weed seeds, thus inhibiting their germination and growth. Fast-growing cover crops like buckwheat or rye can outcompete weeds, reducing the need for herbicides.
4. **Improvement of Soil Structure:** Cover crops, especially those with deep root systems like radishes, break up compacted soil layers and improve aeration and water infiltration. This helps reduce soil compaction and allows the soil to better support future crops.
5. **Water Retention and Carbon Sequestration:** By adding organic matter, cover crops improve the soil's ability to retain water, reducing the need for irrigation. Additionally, the biomass of cover crops helps sequester carbon from the atmosphere, contributing to climate change mitigation.

3 Crop rotation is an agricultural practice where different crops are grown sequentially on the same piece of land over several seasons or years. It involves changing the types of crops planted in a field, rather than cultivating the same crop year after year (monocropping). Crop rotation is a key element in natural farming, regenerative agriculture, and sustainable farming systems, offering numerous benefits that improve soil health, increase biodiversity, and reduce reliance on synthetic inputs.

How Crop Rotation Works

In crop rotation, crops are selected based on their unique characteristics and effects on soil, pests, and nutrient cycling. The rotation schedule is typically designed to balance the needs of the soil and optimize its productivity. For example:

1. **Leguminous Crops (e.g., beans, peas, lentils):** These crops fix nitrogen in the soil by forming symbiotic relationships with nitrogen-fixing bacteria in their root nodules. This process enriches the soil with nitrogen, an essential nutrient for plants.
2. **Heavy Feeders (e.g., corn, wheat):** These crops consume large amounts of soil nutrients, particularly nitrogen, phosphorus, and potassium. After growing a heavy feeder, rotating with a legume or green manure crop can help replenish nutrients.
3. **Root Crops (e.g., carrots, radishes):** Root crops help break up compacted soil and improve its structure, making it easier for subsequent crops to establish healthy root systems.
4. **Cover Crops (e.g., rye, clover):** These crops can be grown between main crops to protect the soil, suppress weeds, and add organic matter through decomposition.

Benefits of Crop Rotation

1. **Soil Fertility Improvement:** Growing the same crop repeatedly can deplete specific nutrients from the soil. In crop rotation, planting nitrogen-fixing crops (like legumes) after nutrient-exhausting crops (like cereals) helps maintain nutrient balance. This reduces the need for chemical fertilizers by allowing the soil to naturally replenish essential nutrients.
2. **Pest and Disease Control:** Pests and pathogens often target specific crops. If the same crop is planted continuously, these pests can build up in the soil, increasing the risk of infestation. By rotating crops, especially with plants that pests or diseases do not favor, their life cycles are disrupted, leading to lower pest and disease pressures.
3. **Weed Management:** Different crops have different growth habits, root structures, and canopy densities, which can suppress different types of weeds. For instance, fast-growing cover crops or cereals can shade out weeds, reducing their establishment and growth. Rotation also helps avoid the dominance of weed species adapted to one type of crop system.

4. **Soil Structure and Health:** Rotating crops with different root structures (e.g., deep-rooted crops like sunflowers and shallow-rooted crops like lettuce) helps improve soil aeration and prevents compaction. Some crops, like legumes, also add organic matter to the soil when their roots and plant residues decompose, enhancing soil structure and microbial activity.
5. **Biodiversity Enhancement:** Crop rotation increases biodiversity both above and below the soil. Growing a variety of crops supports different types of organisms, such as beneficial insects, soil microbes, and pollinators. This contributes to a healthier, more resilient ecosystem.
6. **Reduced Soil Erosion:** Alternating between crops that provide good ground cover (e.g., cereals or cover crops) and those that leave the soil more exposed can help reduce erosion by maintaining better soil coverage throughout the year.

Typical Crop Rotation Cycle

A typical crop rotation might follow a 3- to 4-year cycle

1. **Year 1:** Legume (to fix nitrogen)
2. **Year 2:** Cereal (to take advantage of nitrogen left by the legume)
3. **Year 3:** Root crop or vegetable (to break pest cycles and improve soil structure)
4. **Year 4:** Cover crop or fallow period (to rest the soil and add organic matter)

4 Hand weeding and mechanical weeding are traditional and sustainable methods used to control weeds in farming without the use of herbicides or chemicals. Both techniques involve the physical removal of weeds, but they differ in terms of scale, labor intensity, and the tools used.

Hand Weeding

Hand weeding is the manual removal of weeds using hands or simple tools like a hoe or sickle. This method has been used for centuries, particularly on small farms, home gardens, or areas where precision is required.

How Hand Weeding Works

- **Manual Removal:** Weeds are pulled out directly from the soil by hand, ensuring that the roots are also removed to prevent regrowth. This method is particularly useful for removing perennial weeds, which have deep roots.

- **Selective Weeding:** Hand weeding allows for targeted removal of specific weeds while minimizing damage to the crops. Farmers or gardeners can focus on areas where weed pressure is highest or where invasive species are present.

Benefits of Hand Weeding

1. **Precision:** Hand weeding offers high precision, making it ideal for delicate crops or areas where the proximity of crops and weeds is tight.
2. **Effective for Deep-rooted Weeds:** It is particularly useful for perennial weeds that grow deep into the soil. By manually removing the entire plant, the risk of regrowth is reduced.
3. **No Soil Disturbance:** In some cases, hand weeding can be less disruptive to soil structure than mechanical methods, preserving soil health and reducing erosion.
4. **Cost-effective on Small Scale:** For small farms and gardens, hand weeding can be an economical option since it doesn't require investment in tools or machines.

Limitations of Hand Weeding

- **Labor-intensive:** Hand weeding is physically demanding and time-consuming, especially on large farms. It may require a significant amount of manual labor, making it less efficient for large-scale agriculture.
- **Not Suitable for Large Areas:** Due to its labor intensity, hand weeding is not practical for extensive fields or commercial farming on a large scale.

Mechanical Weeding

Mechanical weeding involves the use of equipment or machinery to remove weeds. This method is often used on larger farms where manual weeding would be too labor-intensive and time-consuming.

How Mechanical Weeding Works

- **Weeding Tools and Equipment:** Mechanical weeding uses implements like hoes, cultivators, and harrows attached to tractors, or hand-pushed mechanical weeders for smaller-scale use. These tools uproot weeds by cutting or disturbing the top layer of the soil.
- **Shallow Tillage:** Mechanical weeders disturb the soil around the crops to dislodge weeds while avoiding damage to the crop plants. Shallow tillage ensures that weed roots are exposed and dry out, preventing regrowth.

Benefits of Mechanical Weeding

1. **Efficiency on Large Scale:** Mechanical weeding is much faster and more efficient for larger areas compared to hand weeding, making it suitable for commercial farming.
2. **Reduced Labor Costs:** Using machinery reduces the need for manual labor, lowering the costs associated with weed control on large farms.
3. **Time-saving:** Mechanical weeding is quicker, especially when done at the right time (before weeds become too large), allowing farmers to manage weed infestations more effectively during peak growing seasons.
4. **Improved Soil Aeration:** Some mechanical weeders also till the soil, which can improve soil aeration and water infiltration.

Limitations of Mechanical Weeding

- **Soil Disturbance:** Mechanical weeding can disturb the soil, which may promote the germination of new weed seeds brought to the surface. It can also lead to soil erosion if done excessively or in areas prone to erosion.
- **Equipment Costs:** The cost of purchasing and maintaining mechanical weeding equipment can be high, especially for small or medium-sized farms.
- **Crop Damage Risk:** If not used carefully, mechanical weeding tools can damage crops, particularly in closely planted fields or with shallow-rooted crops.

5. Zero-Tillage (also known as no-till) and **reduced tillage** are conservation farming practices that minimize or completely avoid disturbing the soil during the planting process. Both methods are integral to sustainable and regenerative agriculture as they aim to improve soil health, enhance biodiversity, and reduce environmental impacts.

Zero-Tillage (No-Till Farming)

Zero-tillage refers to a farming practice in which the soil is left undisturbed from harvest to planting, except for minimal soil disruption during seeding. In this method, seeds are directly planted into undisturbed soil using specialized equipment, without turning or plowing the soil.

How Zero-Tillage Works

- **Direct Seeding:** Seeds are placed into the soil using no-till seed drills or planters, which create narrow slits or holes just big enough for the seed. There is no plowing or deep tillage involved, keeping the soil structure intact.
- **Crop Residue Cover:** The crop residue (like straw or plant stalks) from the previous crop is left on the surface as a natural mulch, protecting the soil and reducing the growth of weeds.

Benefits of Zero-Tillage

1. **Soil Structure Preservation:** No-till farming maintains the natural structure of the soil, reducing compaction and preserving soil layers. Healthy soil structure improves water infiltration and root development.
2. **Soil Erosion Control:** By keeping the soil covered with crop residues, no-till farming prevents erosion from wind and water. This is particularly important in areas prone to heavy rainfall or strong winds.
3. **Water Conservation:** Zero-tillage helps the soil retain moisture by reducing evaporation and improving water infiltration. The crop residues on the surface act as mulch, keeping the soil cool and moist.
4. **Increased Organic Matter:** Crop residues left on the surface break down over time, adding organic matter to the soil. This enhances soil fertility and boosts microbial activity.
5. **Reduced Fuel and Labor Costs:** Since there is no need for ploughing or tilling, farmers save on fuel, labour, and equipment costs. This can also reduce the carbon footprint of farming operations.
6. **Carbon Sequestration:** Zero-tillage practices store carbon in the soil by minimizing the disturbance that would otherwise release stored carbon back into the atmosphere.

Challenges of Zero-Tillage

- **Weed Pressure:** In the absence of tillage, weed control becomes a challenge, and farmers may need to rely on cover crops, mulching, or mechanical weeders to manage weed growth.
- **Slow Transition:** Soils accustomed to conventional tillage may take time to adapt to no-till systems. The benefits of zero-tillage, such as improved soil structure and organic matter, may not be immediate.
- **Equipment Costs:** No-till seed drills and other specialized machinery can require a significant initial investment, although the long-term savings in labour and fuel often offset these costs.

Reduced Tillage

Reduced tillage is a compromise between conventional tillage and no-till farming. It involves limited soil disturbance, often only in the top few inches, and reduces the frequency and intensity of tillage compared to traditional ploughing methods.

How Reduced Tillage Works

- **Shallow Tillage:** Instead of deep ploughing, reduced tillage involves using tools like shallow Plows or cultivators that disturb only the top layer of soil. This minimizes soil disturbance while still allowing for some soil aeration and weed control.
- **Residue Management:** Similar to no-till, reduced tillage systems often leave crop residues on the surface to protect the soil, reduce erosion, and add organic matter.

Benefits of Reduced Tillage

1. **Soil Health Improvement:** While it disturbs the soil less than conventional tillage, reduced tillage still helps preserve soil structure and supports microbial activity.
2. **Reduced Soil Erosion:** Like zero-tillage, reduced tillage helps protect the soil from erosion by leaving plant residues on the surface and minimizing deep soil disruption.
3. **Lower Input Costs:** Reduced tillage requires less fuel, labour, and equipment wear compared to conventional tillage, helping to reduce operational costs.
4. **Weed Control Flexibility:** Reduced tillage provides some flexibility for weed control by allowing shallow tilling to manage weed populations without completely upturning the soil.

Challenges of Reduced Tillage

- **Soil Compaction:** While reduced tillage minimizes deep disturbance, repeated shallow tilling can lead to compaction in the top layer of the soil, particularly if done frequently.
- **Weed Management:** Reduced tillage, like zero-tillage, can result in increased weed pressure, requiring additional strategies like cover cropping or mechanical weeding to manage.

Comparison of Zero-Tillage and Reduced Tillage

- **Soil Disturbance:** Zero-tillage completely avoids soil disturbance, whereas reduced tillage allows for shallow tilling. Both methods aim to limit the adverse effects of conventional tillage, but zero-tillage is the most conservative approach.
- **Weed Management:** Zero-tillage may rely more on natural methods like mulching, cover crops, or organic herbicides for weed control, while reduced tillage may still incorporate shallow tilling for weed suppression.
- **Soil Health:** Both systems improve soil health over time by preserving organic matter and enhancing microbial activity, but zero-tillage is generally more beneficial for long-term soil structure preservation.

[6] Zero-tillage (also known as no-till) and reduced tillage are conservation farming practices that minimize or completely avoid disturbing the soil during the planting process. Both methods are integral to sustainable and regenerative agriculture as they aim to improve soil health, enhance biodiversity, and reduce environmental impacts.

Zero-Tillage (No-Till Farming)

Zero-tillage refers to a farming practice in which the soil is left undisturbed from harvest to planting, except for minimal soil disruption during seeding. In this method, seeds are directly planted into undisturbed soil using specialized equipment, without turning or ploughing the soil.

How Zero-Tillage Works

- **Direct Seeding:** Seeds are placed into the soil using no-till seed drills or planters, which create narrow slits or holes just big enough for the seed. There is no ploughing or deep tillage involved, keeping the soil structure intact.
- **Crop Residue Cover:** The crop residue (like straw or plant stalks) from the previous crop is left on the surface as a natural mulch, protecting the soil and reducing the growth of weeds.

Benefits of Zero-Tillage

1. **Soil Structure Preservation:** No-till farming maintains the natural structure of the soil, reducing compaction and preserving soil layers. Healthy soil structure improves water infiltration and root development.

2. **Soil Erosion Control:** By keeping the soil covered with crop residues, no-till farming prevents erosion from wind and water. This is particularly important in areas prone to heavy rainfall or strong winds.
3. **Water Conservation:** Zero-tillage helps the soil retain moisture by reducing evaporation and improving water infiltration. The crop residues on the surface act as mulch, keeping the soil cool and moist.
4. **Increased Organic Matter:** Crop residues left on the surface break down over time, adding organic matter to the soil. This enhances soil fertility and boosts microbial activity.
5. **Reduced Fuel and Labor Costs:** Since there is no need for ploughing or tilling, farmers save on fuel, labour, and equipment costs. This can also reduce the carbon footprint of farming operations.
6. **Carbon Sequestration:** Zero-tillage practices store carbon in the soil by minimizing the disturbance that would otherwise release stored carbon back into the atmosphere.

Challenges of Zero-Tillage

- **Weed Pressure:** In the absence of tillage, weed control becomes a challenge, and farmers may need to rely on cover crops, mulching, or mechanical weeders to manage weed growth.
- **Slow Transition:** Soils accustomed to conventional tillage may take time to adapt to no-till systems. The benefits of zero-tillage, such as improved soil structure and organic matter, may not be immediate.
- **Equipment Costs:** No-till seed drills and other specialized machinery can require a significant initial investment, although the long-term savings in labour and fuel often offset these costs.

Reduced Tillage

Reduced tillage is a compromise between conventional tillage and no-till farming. It involves limited soil disturbance, often only in the top few inches, and reduces the frequency and intensity of tillage compared to traditional ploughing methods.

How Reduced Tillage Works

- **Shallow Tillage:** Instead of deep ploughing, reduced tillage involves using tools like shallow Plows or cultivators that disturb only the top layer of soil. This minimizes soil disturbance while still allowing for some soil aeration and weed control.

- **Residue Management:** Similar to no-till, reduced tillage systems often leave crop residues on the surface to protect the soil, reduce erosion, and add organic matter.

Benefits of Reduced Tillage

1. **Soil Health Improvement:** While it disturbs the soil less than conventional tillage, reduced tillage still helps preserve soil structure and supports microbial activity.
2. **Reduced Soil Erosion:** Like zero-tillage, reduced tillage helps protect the soil from erosion by leaving plant residues on the surface and minimizing deep soil disruption.
3. **Lower Input Costs:** Reduced tillage requires less fuel, labour, and equipment wear compared to conventional tillage, helping to reduce operational costs.
4. **Weed Control Flexibility:** Reduced tillage provides some flexibility for weed control by allowing shallow tilling to manage weed populations without completely upturning the soil.

Challenges of Reduced Tillage

- **Soil Compaction:** While reduced tillage minimizes deep disturbance, repeated shallow tilling can lead to compaction in the top layer of the soil, particularly if done frequently.
- **Weed Management:** Reduced tillage, like zero-tillage, can result in increased weed pressure, requiring additional strategies like cover cropping or mechanical weeding to manage.

Comparison of Zero-Tillage and Reduced Tillage

- **Soil Disturbance:** Zero-tillage completely avoids soil disturbance, whereas reduced tillage allows for shallow tilling. Both methods aim to limit the adverse effects of conventional tillage, but zero-tillage is the most conservative approach.
- **Weed Management:** Zero-tillage may rely more on natural methods like mulching, cover crops, or organic herbicides for weed control, while reduced tillage may still incorporate shallow tilling for weed suppression.
- **Soil Health:** Both systems improve soil health over time by preserving organic matter and enhancing microbial activity, but zero-tillage is generally more beneficial for long-term soil structure preservation.

7. Intercropping is an agricultural practice where two or more crops are grown together in the same field at the same time, either in alternating rows, strips, or a mixed planting system. This technique is commonly used in sustainable, organic, and natural farming systems to optimize the use of available resources, improve crop productivity, and enhance biodiversity. Intercropping can take various forms, depending on the objectives of the farmer, the types of crops being grown, and the local growing conditions.

Types of Intercropping

1. **Row Intercropping:** In this system, two or more crops are planted in distinct rows. For example, maize and beans might be grown in alternating rows. This arrangement allows for easy management of individual crops while maximizing the complementary benefits of each.
2. **Strip Intercropping:** Similar to row intercropping but on a larger scale, strip intercropping involves planting strips of different crops across a field. Each strip might be several rows wide, and the crops in different strips are often chosen to complement one another, such as corn and alfalfa.
3. **Mixed Intercropping:** In mixed intercropping, two or more crops are planted together without any specific row or strip arrangement. This is often seen in traditional, small-scale farming, where crops are grown closely together in a more natural, intermingled fashion.
4. **Relay Intercropping:** In relay intercropping, one crop is planted first and, before it is fully harvested, another crop is sown. This system staggers the growing seasons, ensuring that there is always a crop growing in the field, maximizing productivity and efficient land use.

Benefits of Intercropping

1. **Efficient Resource Utilization:** Different crops have varying nutrient, water, and light requirements. By growing complementary crops together, intercropping maximizes the efficient use of available resources. For example, deep-rooted crops like maize can draw water and nutrients from deeper soil layers, while shallow-rooted crops like beans or spinach make use of the upper layers of soil.
2. **Weed Suppression:** Intercropping can help suppress weed growth by covering the soil more completely and creating competition for light, nutrients, and space. When fast-growing crops or ground-cover plants are included, they can quickly shade out weeds, reducing the need for manual or chemical weed control.

3. **Pest and Disease Control:** Intercropping increases biodiversity in the field, making it harder for pests or diseases to spread. Pests that target specific crops are less likely to thrive in a diverse planting environment. Certain combinations of crops can also repel pests; for instance, marigolds are often intercropped with vegetables to deter insect pests. Additionally, some plants can serve as trap crops, attracting pests away from the main crop.
4. **Improved Soil Health:** Intercropping contributes to better soil health by increasing organic matter, promoting microbial activity, and reducing soil erosion. Crops with different root systems create more aeration and prevent soil compaction. Legumes, when intercropped, fix nitrogen in the soil, enhancing fertility for other crops.
5. **Risk Diversification:** Intercropping reduces the risk of complete crop failure by growing multiple crops at once. If one crop is affected by disease, pests, or unfavourable weather, the others may still thrive, providing some security for farmers. This is particularly important in regions where environmental conditions can be unpredictable.
6. **Increased Yield and Land Productivity:** Properly managed intercropping systems can lead to higher overall productivity per unit area compared to monocropping. The complementary nature of the crops allows for more efficient use of sunlight, water, and nutrients, often resulting in higher total yields.
7. **Biodiversity and Ecosystem Services:** By promoting biodiversity both above and below the soil surface, intercropping enhances ecosystem services such as pollination, nutrient cycling, and natural pest control. This supports more resilient farming systems that can better withstand environmental challenges.

Challenges of Intercropping

1. **Management Complexity:** Intercropping requires careful planning and management to ensure that the crops chosen complement each other well in terms of growth habits, nutrient needs, and timing. It can also make mechanized farming more challenging, as different crops may have different planting, cultivation, and harvesting times.
2. **Competition Between Crops:** If the crops selected are not well-matched, they may compete for resources like water, light, and nutrients. For example, if both crops have similar root depths or growth periods, one may outcompete the other, reducing the overall yield.

3. **Pest and Disease Spread:** While intercropping can reduce pest and disease pressure, in some cases, pests or diseases may spread from one crop to another, especially if they affect multiple species. Careful crop selection is required to mitigate this risk.

Examples of Common Intercropping Systems

1. **Maize and Legumes (Beans or Peas):** This is one of the most common intercropping systems used in traditional and modern farming. Maize, a tall and nutrient-demanding crop, is paired with legumes, which fix nitrogen in the soil, thus enhancing the nutrient availability for the maize. The maize also provides shade and support for the climbing legumes.
2. **Tomatoes and Basil:** In this system, tomatoes benefit from the presence of basil, which can repel pests like aphids and tomato hornworms. Both crops have different root depths, minimizing competition for nutrients.
3. **Rice and Fish:** This unique form of intercropping, known as integrated rice-fish farming, involves raising fish in flooded rice paddies. The fish provide natural pest control and help fertilize the rice, while rice provides habitat and food for the fish.

8. The stale seedbed technique is a weed management practice used in agriculture to reduce weed pressure before planting a main crop. This technique involves preparing the seedbed (the soil surface where seeds will be planted) well in advance of sowing, allowing weed seeds to germinate naturally. Once the weeds emerge, they are eliminated through various methods before the main crop is planted. The goal is to significantly reduce the weed seed bank in the soil, ensuring fewer weeds compete with the crop during its early growth stages.

How the Stale Seedbed Technique Works

1. **Early Soil Preparation:** The field is tilled or otherwise prepared for planting well before the main crop is sown. This creates an ideal environment for weed seeds in the soil to germinate. The soil may be tilled shallowly to disturb the weed seeds and bring them closer to the surface, where they can sprout.
2. **Weed Seed Germination:** After the seedbed is prepared, the farmer waits for weed seeds to germinate. Since many weed species germinate quickly under favourable conditions, this process can happen within days to weeks, depending on the weed species and environmental conditions.
3. **Weed Elimination:** Once the weeds have germinated and emerged above the soil, they are destroyed using one of the following methods:

- **Shallow Tillage:** A light tilling or harrowing can be done to uproot young weeds without disturbing the seedbed too much.
- **Flaming:** A propane-fuelled flame weeder is used to scorch the weeds, killing them without disturbing the soil.
- **Herbicides:** In some farming systems, a non-residual herbicide can be applied to kill the weeds. This is more common in conventional farming, but organic systems avoid herbicides.
- **Hand or Mechanical Weeding:** Depending on the scale of the farm, hand weeding or the use of mechanical weeders may also be used.

4. **Main Crop Planting:** After the weeds are cleared, the main crop is planted into the now "stale" seedbed. Since most of the weed seeds have already germinated and been killed, the main crop faces less competition from weeds, allowing it to establish more easily.

Benefits of the Stale Seedbed Technique

1. **Reduced Weed Pressure:** The stale seedbed technique allows farmers to control a significant portion of the weeds before the main crop is even planted. This reduces weed competition during the crop's critical early growth stages, leading to better crop establishment and potentially higher yields.
2. **Less Reliance on Herbicides:** In systems aiming to reduce or eliminate chemical inputs, the stale seedbed technique offers a non-chemical method of weed control. By removing weeds mechanically or thermally before planting, farmers can reduce the need for herbicide applications during the growing season.
3. **Preservation of Soil Structure:** Since the seedbed is prepared well in advance and the subsequent weed control methods are typically shallow or non-disruptive, the stale seedbed technique helps preserve soil structure. Minimizing deep tillage protects soil health and reduces erosion risks.
4. **Time and Labor Efficiency:** Once the weeds are killed, the need for post-planting weeding is reduced, saving time and labour during the crop's growth period. Early weed control also minimizes competition for water, nutrients, and light.
5. **Enhanced Crop Establishment:** By reducing weed competition early on, crops are able to establish strong roots and grow without being overshadowed by fast-growing weeds. This leads to better crop health and higher productivity.

Challenges of the Stale Seedbed Technique

1. **Timing Sensitivity:** The technique requires careful timing to ensure that enough weeds have germinated before elimination, but not so much time that they grow large and become harder to control. Weather conditions like rain or drought can affect the timing and success of the process.
2. **Requires Extra Time Before Planting:** Farmers need to plan and allow additional time between soil preparation and planting to give weeds time to germinate and be eliminated. This may not always be feasible in tight planting schedules or regions with short growing seasons.
3. **Not Suitable for All Weed Types:** The stale seedbed technique is most effective against weed species that germinate quickly after tillage. Some weed species may not germinate during the preparation phase or may emerge later, requiring additional post-planting weed control measures.
4. **Labor and Equipment Costs:** Depending on the weed control method used (e.g., flaming, mechanical weeding), the initial costs of equipment or labour can be high, particularly for small-scale farmers without access to mechanized equipment.

Applications of the Stale Seedbed Technique

- **Organic Farming:** Since the stale seedbed technique can minimize or eliminate the need for herbicides, it is widely used in organic farming systems to manage weeds naturally.
- **Vegetable and Row Crops:** This technique is commonly used in high-value vegetable crops like carrots, lettuce, onions, and others that are sensitive to early weed competition.
- **Reduced Tillage Systems:** The stale seedbed method works well in reduced tillage systems where the goal is to minimize soil disturbance and improve soil health while managing weeds effectively.

9. Allelopathy is a biological phenomenon where certain plants release chemicals into the environment that influence the growth, survival, and reproduction of other plants. These chemicals, called allelochemicals, can either inhibit or promote the growth of neighbouring plants, although allelopathy is more often associated with suppressing competition. This process plays an important role in natural ecosystems and agricultural systems, influencing plant interactions, weed control, and soil health.

How Allelopathy Works

Plants release allelochemicals into their surroundings through various means:

- **Root Exudation:** Allelochemicals are secreted from plant roots into the soil.
- **Leaf Litter and Decomposition:** When leaves, stems, and other plant parts fall to the ground and decompose, they release allelopathic chemicals into the soil.
- **Volatile Compounds:** Some plants release chemicals into the air, which then settle onto the surrounding soil or plants.
- **Leaching from Rain or Dew:** Allelochemicals can be washed off plant surfaces by rain or dew and enter the soil.

Once these allelochemicals are released, they can affect neighbouring plants in a variety of ways, such as:

- **Inhibiting Germination:** Certain plants can prevent the seeds of other species from germinating by releasing chemicals that inhibit seed growth.
- **Stunting Growth:** Allelochemicals may interfere with the root or shoot growth of nearby plants, slowing their development or even killing them.
- **Disrupting Nutrient Uptake:** Some allelochemicals can affect a plant's ability to absorb essential nutrients from the soil, putting it at a competitive disadvantage.

Examples of Allelopathic Plants

1. **Black Walnut (Juglans nigra):** One of the most well-known examples of allelopathy is the black walnut tree, which releases a compound called juglone from its roots, leaves, and fruit. Juglone inhibits the growth of many plants, including tomatoes, apples, and some grasses, making it difficult to grow certain crops or plants near black walnut trees.
2. **Sorghum (Sorghum bicolor):** Sorghum is often used in agriculture for its allelopathic properties. It releases chemicals called sorgolactones, which can suppress the germination and growth of weeds. Sorghum is sometimes used as a cover crop to reduce weed pressure in subsequent planting cycles.
3. **Sunflowers (Helianthus annuus):** Sunflowers are allelopathic and release chemicals from their roots and decaying plant matter that can inhibit the growth of other plants, especially grasses and legumes.

Farmers may experience poor growth of crops planted after sunflowers due to the allelopathic residues left in the soil.

4. **Eucalyptus (Eucalyptus spp.):** Eucalyptus trees release volatile oils and other allelochemicals that can suppress the growth of understory vegetation, which helps the tree maintain dominance in its environment. These oils can also impact the germination of nearby plants.

Benefits of Allelopathy in Agriculture

1. **Natural Weed Control:** Allelopathic plants can be used in crop rotation or as cover crops to suppress weeds naturally, reducing the need for herbicides. For example, planting rye (Secale cereale) as a cover crop can inhibit the growth of weed seeds, leading to less competition for subsequent crops.
2. **Improved Crop Yield:** By reducing competition from weeds, allelopathic plants can help increase the yield of main crops. Some allelopathic species, such as sorghum and barley, are grown in rotation to keep weed populations low and boost soil fertility for future crops.
3. **Reduced Pesticide Use:** In some cases, allelopathy can help manage pests indirectly. For example, by suppressing certain weed species that host pests, allelopathic plants contribute to reducing the pest population, lowering the need for chemical pesticides.
4. **Soil Health and Conservation:** Many allelopathic plants, especially cover crops like rye and clover, contribute organic matter to the soil when they decompose. This improves soil structure, water retention, and nutrient availability while maintaining biological diversity in the soil.

Challenges of Allelopathy in Agriculture

1. **Unintended Crop Suppression:** While allelopathic plants can suppress weeds, they can also inadvertently affect the growth of beneficial crops. For example, residues from allelopathic cover crops might negatively impact the germination of subsequent crops if not managed carefully.
2. **Soil Residue Persistence:** Some allelochemicals remain in the soil for long periods, which can delay or reduce the growth of crops planted after an allelopathic species. Farmers need to be aware of this when planning crop rotations or planting schedules.
3. **Complex Interactions:** Allelopathy is influenced by various environmental factors, such as soil type, moisture levels, and microbial activity, which can make it difficult to predict the effectiveness or impact of allelopathic plants in specific conditions.

4. **Limited Research and Knowledge:** While the effects of allelopathy are well-documented for certain plants, the broader application of this knowledge in farming systems is still developing. More research is needed to better understand which plant combinations work best and how to manage allelopathic interactions for optimal crop production.

9. Natural herbicides, often referred to as bio-formulations or organic herbicides, are plant-based or naturally derived substances used to control unwanted weeds in agricultural and horticultural systems. Unlike synthetic herbicides, natural herbicides are typically considered safer for the environment, human health, and beneficial organisms, making them a popular choice in organic farming and sustainable agriculture practices. They can be derived from various sources, including plants, microorganisms, and minerals, and can be effective in managing weeds with fewer negative impacts compared to conventional chemical herbicides.

Types of Natural Herbicides

1. **Plant Extracts:** Many natural herbicides are made from extracts of specific plants that possess allelopathic properties. These plants release chemicals that inhibit the germination and growth of other plants. Examples include:
   - **Corn Gluten Meal:** This byproduct of corn processing contains compounds that prevent seed germination and can be effective against annual weeds.
   - **Vinegar (Acetic Acid):** Household vinegar, containing acetic acid, can kill young weeds when sprayed directly on them. It works by dehydrating the plant tissues.
2. **Essential Oils:** Essential oils from various plants have been shown to possess herbicidal properties. Oils such as clove oil, cinnamon oil, and rosemary oil can disrupt plant cell membranes and inhibit growth. They are often used in concentrated forms and may require multiple applications for effective weed control.
3. **Microbial Herbicides:** Certain bacteria and fungi can act as natural herbicides by infecting and killing specific weed species. For example, Phoma macrostoma and Myrothecium verrucaria are fungal pathogens that can be used to target particular weeds without harming crops.
4. **Mineral-Based Herbicides:** Some natural herbicides are derived from minerals. For instance, sodium chloride (salt) can be used in small doses to desiccate weeds. However, excessive use can harm soil health and surrounding plants.

5. **Biopesticides:** Some biopesticides, which are primarily used for pest control, can also have herbicidal effects. These products may contain naturally occurring substances that deter weed growth.

Mechanisms of Action

Natural herbicides work through various mechanisms

- **Desiccation:** Many natural herbicides cause dehydration in plants, leading to their death. For example, vinegar works by drying out the plant tissues.
- **Inhibition of Germination:** Some natural herbicides prevent seeds from germinating by affecting the processes necessary for seedling establishment.
- **Allelopathy:** Certain natural herbicides release chemicals that interfere with the growth and development of surrounding plants, suppressing their growth.
- **Pathogenic Infection:** Microbial herbicides can infect and kill weeds, thereby reducing competition for crops.

Advantages of Natural Herbicides

1. **Environmental Safety:** Natural herbicides are generally less toxic to humans, animals, and beneficial organisms, making them a safer choice for sustainable agriculture.
2. **Reduced Chemical Residues:** They do not leave harmful chemical residues in the soil or on crops, aligning with organic farming standards and consumer preferences for chemical-free produce.
3. **Biodiversity Promotion:** Using natural herbicides can support a more biodiverse ecosystem by minimizing negative impacts on non-target plants and beneficial organisms, such as pollinators and soil microbes.
4. **Soil Health:** Many natural herbicides do not adversely affect soil health, maintaining the soil's structure, nutrient content, and microbial activity, which are crucial for long-term agricultural sustainability.
5. **Compliance with Organic Standards:** Natural herbicides are often approved for use in organic farming, making them essential tools for organic farmers seeking effective weed management strategies.

Challenges of Natural Herbicides

1. **Efficacy:** Natural herbicides may not be as effective as synthetic herbicides, especially for established weeds or those with deep root systems. They often require multiple applications and may work best on young, actively growing weeds.
2. **Application Timing and Conditions:** The effectiveness of natural herbicides can be influenced by environmental conditions such as temperature, humidity, and soil moisture. Farmers must carefully time applications for optimal results.
3. **Cost and Availability:** Some natural herbicides can be more expensive to produce and apply than conventional options. Limited availability in some regions may also pose challenges for farmers.
4. **Short Residual Activity:** Many natural herbicides have a shorter residual effect, meaning they may require more frequent applications to maintain weed control.
5. **Knowledge and Expertise:** Effective use of natural herbicides often requires more knowledge about plant biology, weed management, and specific application techniques compared to conventional herbicides.

Examples of Natural Herbicides

1. **Corn Gluten Meal:** This pre-emergent herbicide prevents the germination of weed seeds and is often used in lawns and gardens.
2. **Vinegar (Acetic Acid):** A common household product that acts as a non-selective herbicide, particularly effective against annual weeds.
3. **Clove Oil:** Known for its strong antimicrobial properties, clove oil can also kill young weeds when applied directly.
4. **Essential Oil Blends:** Commercially available natural herbicides often contain a combination of essential oils for enhanced effectiveness against a range of weed species.

10. Livestock integration, often referred to as **grazing integration** or **mixed farming**, is an agricultural practice that involves the strategic inclusion of livestock within crop production systems. This practice is gaining traction as a sustainable approach to farming, offering numerous environmental, economic, and agronomic benefits. The integration of livestock into crop production systems helps optimize land use, enhances soil fertility, and contributes to pest and weed management while improving overall farm resilience.

Key Concepts of Livestock Integration

1. **Nutrient Cycling:** Livestock play a vital role in nutrient cycling on farms. Their manure serves as a natural fertilizer, returning essential nutrients like nitrogen, phosphorus, and potassium to the soil. This reduces the need for synthetic fertilizers, enhancing soil fertility and promoting healthier crop growth.
2. **Weed and Pest Management:** Grazing animals can help control weeds and pests naturally. By grazing on specific weed species, livestock reduce their prevalence and limit their spread. This biological control method can decrease the reliance on herbicides and pesticides, promoting a more sustainable approach to pest management.
3. **Soil Health Improvement:** Livestock integration contributes to improved soil structure and health. Grazing animals naturally aerate the soil through their movement and trampling, which enhances water infiltration and reduces soil compaction. Additionally, manure adds organic matter to the soil, fostering beneficial microbial activity and improving overall soil health.
4. **Diversity in Farming Systems:** Integrating livestock into cropping systems adds diversity, which can enhance farm resilience against climate fluctuations, pests, and diseases. This diversification can stabilize farm income and reduce financial risks by allowing farmers to produce both crops and livestock.
5. **Utilization of Marginal Lands:** Livestock can graze on land that may not be suitable for crop production, such as steep slopes or rocky terrains. This allows for more efficient use of available resources and increases overall productivity on the farm.

Methods of Livestock Integration

1. **Rotational Grazing:** In this method, livestock are moved between different pastures or fields, allowing grazed areas to recover while providing nutrients through manure. Rotational grazing prevents overgrazing, encourages plant regrowth, and promotes biodiversity in pasture ecosystems.
2. **Cover Cropping:** Farmers can use cover crops, which are planted specifically to improve soil health and prevent erosion, as forage for livestock. Grazing cover crops can provide livestock feed while enhancing soil quality, particularly during off-seasons.
3. **Intercropping:** In intercropping systems, crops and livestock are integrated closely in the same field. For example, livestock may graze on

crop residues after harvest, effectively utilizing leftover plant material and minimizing waste.

4. **Agroforestry:** This approach combines trees, crops, and livestock in a single system. Livestock can graze in tree-dominated landscapes, benefiting from shade and shelter while contributing to nutrient cycling and weed control.

Benefits of Livestock Integration

1. **Enhanced Productivity:** The synergy between crops and livestock can lead to higher overall productivity. Livestock provide manure for crops, while crop residues offer feed for livestock, creating a closed-loop system that maximizes resource use.
2. **Economic Benefits:** Integrating livestock can diversify income streams for farmers. By producing both crops and livestock, farmers can mitigate risks associated with market fluctuations in either sector. Additionally, reduced input costs for fertilizers and pest management can improve profitability.
3. **Environmental Sustainability:** Livestock integration supports sustainable farming practices by reducing chemical inputs, enhancing biodiversity, and promoting healthier soils. This contributes to improved ecosystem services, such as carbon sequestration and water conservation.
4. **Improved Animal Welfare:** Grazing livestock in integrated systems often allows for more natural behaviors and healthier living conditions. Access to pasture can improve the well-being of animals, leading to better growth rates and overall health.
5. **Resilience to Climate Change:** Diverse farming systems that include livestock are generally more resilient to climate change impacts. They can better adapt to variations in weather patterns, reducing vulnerability to droughts or floods.

Challenges of Livestock Integration

1. **Management Complexity:** Integrating livestock with crops requires careful planning and management. Farmers must balance the needs of both systems, including grazing schedules, pasture management, and crop rotations.
2. **Disease Management:** The presence of livestock can introduce diseases to crops and vice versa. Farmers need to implement effective biosecurity measures to minimize disease risks and protect both livestock and crops.
3. **Resource Competition:** Livestock and crops may compete for resources

such as water and nutrients. Farmers must manage these interactions carefully to ensure that both systems benefit from integration.

4. **Infrastructure Needs:** Integrating livestock may require additional infrastructure, such as fencing, watering systems, and shelters, which can incur initial costs for farmers.

11. Solarization is an eco-friendly agricultural technique that uses solar energy to control pests, diseases, and weeds in soil. This method involves covering moist soil with clear plastic for several weeks during hot weather to trap solar radiation, effectively heating the soil to temperatures that can kill harmful organisms and seeds. Solarization is particularly beneficial in organic farming systems and is widely used for soil management, improving soil health, and preparing fields for planting.

How Solarization Works

1. **Soil Preparation:** The process begins with preparing the soil by tilling or loosening it to promote uniform heating and moisture retention. This also helps incorporate any organic matter or amendments that may enhance the effectiveness of solarization.
2. **Moistening the Soil:** The soil should be thoroughly moistened before covering it with plastic. Moisture is crucial because it enhances heat retention and helps kill pathogens, pests, and weed seeds more effectively.
3. **Covering with Plastic:** Clear plastic sheeting, typically 1 to 6 mil thick, is spread over the moistened soil. The plastic must be tightly sealed at the edges to prevent heat and moisture loss. Using clear plastic is essential because it allows sunlight to penetrate and heat the soil beneath.
4. **Heating Phase:** The sun's rays penetrate the plastic, heating the soil underneath. Depending on the climate and weather conditions, soil temperatures can reach 60°C (140°F) or higher within a few days. This heat can persist for several weeks, effectively killing weed seeds, pathogens, nematodes, and insect larvae.
5. **Duration:** The solarization period typically lasts between 4 to 8 weeks, depending on the climate, time of year, and desired soil temperatures. In cooler regions, a longer period may be necessary to achieve effective results.
6. **Removal of Plastic:** After the solarization period, the plastic is removed, and the soil is left to cool before planting. The treated soil will generally be free of many pests and diseases, creating a healthier environment for subsequent crops.

Benefits of Solarization

1. **Pest and Disease Control:** Solarization effectively reduces populations of soil-borne pathogens, nematodes, and weed seeds. This leads to healthier crops with reduced risk of disease and pest infestation.
2. **Weed Management:** By heating the soil to lethal temperatures, solarization can significantly reduce weed populations, making it easier to establish new crops without competition.
3. **Soil Health Improvement:** Solarization can improve soil structure and promote beneficial microbial activity. The process helps create a more favorable environment for plant growth and can enhance nutrient availability.
4. **Reduced Chemical Inputs:** This method minimizes the need for chemical herbicides and pesticides, aligning with organic farming practices and reducing environmental impact.
5. **Cost-Effective:** Solarization is a low-cost technique that requires minimal investment compared to traditional chemical treatments or soil fumigation.

Limitations of Solarization

1. **Weather Dependence:** Solarization is most effective in regions with abundant sunlight and warm temperatures. Cloudy or rainy weather can significantly reduce soil heating and diminish the effectiveness of the process.
2. **Duration:** The time required for effective solarization can be a limitation for some farmers, especially if they have a short growing season. Planning is essential to ensure that solarization is completed before the planting season.
3. **Soil Type:** The effectiveness of solarization can vary based on soil type and moisture content. Sandy soils may not retain heat as effectively as clay soils, potentially leading to uneven results.
4. **Limited Control of Deep-Rooted Weeds:** While solarization can kill many weed seeds in the top few inches of soil, it may not be effective against deep-rooted perennial weeds. Additional management strategies may be necessary for complete control.
5. **Temporary Solution:** The benefits of solarization may be short-lived if new pest and weed populations are introduced into the soil after treatment. Sustainable management practices should be implemented to maintain soil health and prevent reinfestation.

Applications of Solarization

1. **Vegetable Production:** Solarization is commonly used in vegetable gardens and commercial crop production to control pests and weeds, ensuring healthier and more productive crops.
2. **Organic Farming:** Organic farmers often rely on solarization as a non-chemical method for managing soil-borne diseases and weeds, supporting sustainable agriculture practices.
3. **Nursery Management:** Nurseries use solarization to prepare potting soils and seedbeds, helping to ensure healthy plants and minimize disease incidence.
4. **Turf and Landscape Management:** Solarization can be applied in turf management and landscaping to eliminate weeds and pests before establishing new lawns or gardens.

6

Insect Pest Management in Natural Farming

The crop production has been found to make wide use of chemicals with hundreds of newly imported molecules. Unnecessary and excessive use of inorganic chemicals have led to an upsurge in the ecosystem and imbalance of nature. Natural enemies and bees get destroyed with chemical pesticides. Other side effects include resurgence of target pests and outbreak of secondary pests. Moreover, the residues of the pesticides in food products and environment have caused grave health issues. The contamination of pesticides into the fruits causes a quality decrease of the fruits, which is widely reported. The best available solution for sustainable agriculture can be the non-chemical organic/ natural farming that does not permit the usage of inorganic and synthetic chemicals for crop cultivation. Natural farming is the ecology-based production management system promoting and It provides biodiversity, biological cycles, and soil biological activity. The basic principles of organic/natural farming are materials and practices that enhance the ecological balance of natural systems that integrate parts of the farming system into the ecosystem. The first purpose of natural farming is to improve the health and productivity of interdependent communities of soil life, plants, animals, and people, that is termed as 'one health concept'.

Natural farming is a system in which laws of nature are applied to agricultural practices. It has an advantage of natural biodiversity on every area of farmed resources; it also fosters the intricacy of living organisms, plant and animal alike, which would form each specific ecosystem to grow in harmony with food crops. Insects are indeed very mobile and well adapted to farm production systems as well as pest control tactics. Organic/natural farms focus on management, not eradication, of these insects. Thus, success in organic/natural farms depends on learning the biological, ecological, and behavioural information of the insects. The biological information as to what the insect needs to survive can be used to determine if insect pests can be deprived of some vital resource. Ecological information can be used to inform how the insect interacts with the environment and other species. shape an environment resistant to pests.

Behavioural information addresses both the pest and beneficial insects, and the manner in which the insect goes about getting what it needs to live can be manipulated to protect the crops. Under the natural farming system, insect pest problems may be managed through cultural, mechanical or physical means; development of habitat for natural enemies of pests and non-synthetic control such as traps, lures and repellents.

Components of pest management under natural farming

The following components may be included in pest management under natural farming system A. Ecology based pest management and habitat diversification

B Use of resistant varieties

Physical methods of pest management

D. Mechanical methods of pest management

E. Use of plant products / botanicals like astras

F. Use of insect pheromones

G. Biological control of pests

H. Indigenous technical knowledge

A. Ecology based pest management and habitat diversification :

1. **Intercropping system :** It has been found as a favorable inter-cropping system in reducing the population and damage caused by many insect pests due to one or more of the following reasons. • Pest outbreak less in mixed stands due to crop diversity than in sole stands • Availability of alternate prey • Decreased colonization and reproduction in pests • Chemical repellency, masking, feeding inhibition by odours from non-host plants. • Physical barrier to plants. Some examples of an intercropping system wherein a decrease in the level of damage was observed are shown in the following table:

Table 1: Effect of intercropping system on pest levels

No.	Crop		Pest reduced	Reference
	Main crop	Intercrop		
1.	Sorghum	Red gram	Earhead bug	Raheja (1973)
2.	Sorghum	Cowpea	*Chilo parte Hus*	Balasubramainian (2000)
3.	Pigeon pea	Sorghum	*Empoasca kerri*	Sekhar efa/(1997)
4.	Green gram	Sorghum	*E. kerri*	Sekhar efa/(1997)
5.	Ground nut	Sorghum	*E. kerri*	Sekhar *et* <?/(1997)
6.	Pigeon pea	Sorghum	*H. armigera*	Mohammed and Rao (1998)

No.	Crop		Pest reduced	Reference
	Main crop	Intercrop		
7.	Chickpea	Wheat/ mustard/ Safflower	*H. armigera*	Das(1998)
8.	Sugarcane	Greengram / Blackgram	Early shoot borer	Rajendran *etai.* (1998)

2. **Trap cropping :** Directly cultivating crops so as to attract or attact insects or other organisms, such as nematodes, for protection of the targeted crop from pests. It is achieved by either Exclude the pests from the crop or Concentrate them in a specific portion of the field where they can be commercially eliminated. Raising of 2 rows of mustard as a trap crop per 25 rows of cabbage is advised for the control of diamond back moth. First mustard crop is sowable15 days before cabbage planting or 20 days old mustard seedlings are planted. Indirect control Direct sowing of castor as a border crop and in irrigation channels acts as an indicator or trap crop for Spodoptera litura. Sowing African tall marigold 40 days old seedlings along with tomato 25 days old seedlings at the ratio of 1:16 results in alternate rows, which minimizes the damage caused by Helicoverpa. These include trap crops like marigold, which attracts pests such as the American bollworm to lay their eggs; barrier crops such as maize/jowar that prevent the migration of sucking pests such as aphids; and guard crops like castor, which attracts Spodoptera litura in cotton fields (Murthy and Venkateshwarulu, 1998).

Table 2: List of successful examples of trap crop

No.	Main Crop	Trap crop	Pest
1.	Tobacco / cotton/ groundnut	Castor	*Spodoptera Hiura*
2.	Maize	Sorghum	Shoot fly, Stem borer
3.	Cotton	Onion / Garlic	*Th rips tabaci*

3. **Fertilizer management :** It depends upon the nutritional status of soil which indirectly affects the pests. High levels of N fertilizer always favor insects and makes plants more susceptible to infestation by insects. Conversely, lower potassium supply favors the insects development, whereas optimum and high K have depressant effects. The table-3 indicates that application of Jeevamrit increases the level of potash in the plants and soil.

Table 3: Effects of host plant nutrition on insect pests

S. No.	Host plant	Insect	Response	Reference
1.	Rice	Thrips, GLH, Whorl maggot, Leaf folder	High K application reduced pest incidence	Subramanian and Balasubramanian (1976)
2.		Leaf folder, gall midge, BPH, Yellow stem borer, WBPH	High N levels increased pest population and damage	Upadhyay ef a/.(1981) Narayanan *et al.* (1973) Saroja and Raju (1981)
3.	Wheat	Cutworm *(Mythimna separata)*	High N increased incidence	Deol *et al.* 1987
4.	Sorghum	Shoot fly	High P reduced incidence	Bangar, 1985
5.	Cotton	Pink boll worm, leafhopper	High N increased incidence	Simwat *et al.* 1987, Purohit and Deshpande, 1992
6.	Chickpea	*HeHcoverpa armigera*	N increased infestation, while, P and K reduced	Yadav, 1987

4. **Planting dates & crop duration:** Planting dates should be so adjusted that the susceptible stage of crop synchronizes with the most inactive period or lowest pest population. The plantings should be also based 91 on information on pest monitoring, as the data varies with location. Crop maturity also plays an important role in pest avoidance. The following table (Table 4) shows the importance of planting dates on pest population and damage.

Table 4: Role of planting dates on pest population and damage

No.	Host plant	Insect	Response	Reference
1.	Rice	Leaf folder	Early planted rice (up to 3rd week of June) suppressed population	Dhaliwal *et al.* (1988)
2.		BPH	Planting during end of July in Kharif and early in Rabi escapes attack in Andhra Pradesh	Krishnaiah *et al.* (1986)
3.		Gall midge	Lowest incidence if planted in August or October	Uthamasamy and Karuppuchamy (1986)
4.	Sorghum	Shoot fly	Advancing sowing date (September to October) decreased incidence	Kotikal and Panchbavi (1991)
5.	Cotton	Leafhopper	Higher incidence in late sown crop	Dhawan *et al.* (1990)

No.	Host plant	Insect	Response	Reference
6.	Chickpea	Pod borer	For every 10 days delay in sowing 4.02% increase in pod damage	Devendra Prasad *et al.* (1989)
7.	Tomato	Whitefly	Incidence is less if planted within July to November	Saikia abd Muniappa (1989)
8.	Chillies	Thrips	Late planted crop severely affected by thrips and leaf curl virus	Bagle (1992)

5. **Planting density :** Plant nutrient status, interplant spacing, canopy structure, etc., affect insect behaviour in searching food, shelter and oviposition site. It also affects natural enemy population. The effect of plant density on pest population is shown in Table 5.

Table 5. Effect of plant density on pest population

S. No.	Crop	Spacing/ density	Insect	Response	Reference
1.	Rice	Dense planting	Leaf folder, BPH	High incidence	Kushwaha and Sharma (1981); Kalode and Krishnaiah (1991)
2.	Chickpea	Dense plant population	*H. armigera*	High incidence	Yadav(1987)
3.		Less plant population	*Aphis craccivora*	High incidence	Lal efa/(1989)

6. **Destruction of alternate host plants :** Many insects use a wide range of cultivated plants especially weeds as alternate hosts for off season carry-over of population. Matteson et al. (1984) reported that weeds around the crop can alter the proportion of harmful and beneficial insects that are present and increase or decrease crop damage.

Table 6: Effect of alternate hosts on pest damage

S. No.	Crop	Pest	Alternate host to be removed	Reference
1.	Groundnut	Th rips	*Achyranthus aspera*	Mohan Daniel *et al.* (1984)
2.	Rice	Gallmidge	Wild rice *(O.nivara)*	Kalode and Krishnaiah (1991)
3.		GLH	*Leersia hexandra*	
4.			*Echinochloa colonum*	
5.			*E.crusgalli*	
6.			*Cynodan dactylon*	
7.		WBPH	*Chleres barbata*	
8.	Sorghum	Earhead midge	Grassy weeds	Prem Kishore (1987)

7. Water management : Availability of water in requisite amount at the appropriate time is crucial for proper growth of crop. Hence, water affects the associated insects by many ways such as nutritional quality and quantity, partitioning of nutrients between vegetative growth and reproduction etc. The following table shows the effect of irrigation on pest population / damage.

Table 7: Effect of irrigation on pest population / damage

No.	Crop	Insect	Response	Reference
1.	Rice	Mealy bug	Continuous stagnation of 5 cm water reduced incidence	Gopalan eta/. (1987)
2.	Rice	Caseworm and BPH	Draining of water to field capacity reduces incidence	Thomas (1986)

8. Crop rotation : Sustainable systems of agricultural production are seen in areas where proper mixtures of crops and varieties are adopted in a given agro-ecosystem. Monocultures and overlapping crop seasons are more prone to severe outbreak of pests and diseases. For example growing rice after groundnut in garden land in puddled condition eliminates white grub.

B. Use of resistance varieties

Using resistant varieties of crops for pest management is an integral part of integrated pest management (IPM) strategies. It involves breeding and planting crop varieties that have natural resistance to certain pests, reducing the reliance on chemical pesticides and other control methods. Here's how resistant varieties contribute to pest management:

1. Reduction in Pesticide use

- Natural resistance lowers the need for chemical pesticides. This minimizes environmental pollution and reduces harmful effects on non-target organisms such as beneficial insects, birds, and pollinators.

2. Cost-Effective

- Farmers save money on pesticide purchases and application costs. In the long term, resistant varieties can also help avoid yield losses, improving profitability.

3. Durable Control

- Resistant varieties offer a long-term solution compared to chemical controls that can lose efficacy as pests evolve resistance to them. Resistant crops can maintain productivity even under pest pressure.

4. Environmentally Friendly

- Since it reduces the reliance on synthetic chemicals, using resistant varieties helps reduce environmental contamination. This is especially beneficial in areas with sensitive ecosystems.

5. Slowing Pest Evolution

- Continuous use of the same pesticide can cause pest populations to evolve resistance. By incorporating pest-resistant varieties into a management strategy, the development of resistant pest strains can be slowed.

6. Target-Specific Resistance

- Many resistant varieties are bred to be species-specific, meaning they target specific pests. This allows for precise management of pest populations without affecting other beneficial species.

Types of Resistance

- **Antibiosis:** Affects pest survival, reproduction, or development.
- **Antixenosis:** Affects pest behavior, deterring them from feeding or reproducing on the plant.
- **Tolerance:** Plants withstand pest damage better, reducing yield loss even when pests are present.

Challenges

- **Pest adaptation:** Over time, pests may overcome resistance traits, requiring continuous breeding efforts to keep up.
- **Limited availability:** In some regions, resistant varieties may not be available for all crops or pests.
- **Potential trade-offs:** Some resistant varieties may have lower yields or other agronomic traits that are less desirable compared to susceptible varieties.

Examples

- **Bt Crops:** Crops like Bt cotton and Bt corn have been genetically modified to express a bacterial protein toxic to certain pests (e.g., the European corn borer).
- Rice varieties resistant to brown planthopper (BPH) through conventional breeding.
- Wheat varieties resistant to rusts, a fungal pest.

C. Physical methods of pest management

Physical methods of pest management are non-chemical techniques used to control or suppress pest populations through direct mechanical or physical actions. These methods are environmentally friendly, can be part of an integrated pest management (IPM) strategy, and often work well in conjunction with biological and cultural controls. Here are some common physical pest management methods:

1. Handpicking

- Involves manually removing pests such as caterpillars, beetles, or large insects from crops or garden plants.
- **Suitable for small-**scale farms, gardens, and for specific pests like larger insects.
- **Example:** Handpicking tomato hornworms off plants.

2. Traps

- **Sticky Traps:** Coated with adhesive to catch flying insects like aphids, whiteflies, or fungus gnats.
- **Pheromone Traps:** Use pheromones (chemical attractants) to lure and trap specific pests, such as moths.
- **Light Traps:** Attract flying insects with light and capture or kill them with electric grids or sticky surfaces.
- **Mechanical Traps:** Snap traps or bucket traps to control larger pests like rodents.

3. Barriers

- Physical barriers prevent pests from reaching plants or entering structures.
- **Examples**
  - Row covers: Lightweight fabric covers placed over crops to block pests like cabbage moths or flea beetles.
  - Mulches: Organic or synthetic mulches can prevent pests like weeds or insects from accessing crops.
  - Netting/Fencing: Prevents birds, deer, or rodents from accessing crops.
  - Copper bands: Placed around plant beds to deter slugs and snails.

4. Soil Solarization

- This involves covering soil with clear plastic sheeting during hot periods to trap solar radiation and raise soil temperature, which kills soil-borne pests, pathogens, and weed seeds.
- Effective against nematodes, soil fungi, and weed seeds in the top few inches of soil.

5. Mechanical Cultivation

- Tilling or cultivating the soil can disrupt pest habitats and destroy insect eggs, larvae, and weed seeds.
- This method is often used to control weeds and root-feeding pests.

6. Water Pressure/Jet Spraying

- High-pressure water sprays can be used to dislodge and remove pests like aphids, spider mites, or scale insects from plant surfaces.
- Non-toxic and safe for plants if used carefully.

7. Heat Treatments

- **Flaming:** Using a flame weeder to pass over crops and destroy small weeds or pests without damaging larger plants.
- Hot Water Treatment: Soaking seeds or seedlings in hot water to kill pathogens and insect pests like nematodes or seed-borne diseases.
- Steam Sterilization: Applying steam to sterilize soil and eliminate pests such as fungi, bacteria, and weed seeds.

8. Cold Treatments

- Exposure to low temperatures can kill certain pests, especially in storage facilities.
- **Example:** Freezing grains to eliminate insect infestations.

9. Vacuuming

- Used in some agricultural practices to physically remove pests such as aphids or spider mites from crops.
- Useful in greenhouses or other controlled environments.

10. Mulching

- Organic or synthetic mulches can prevent insect pests from laying eggs in the soil or can create a barrier against weeds and soil-borne pests.

- Certain mulches, like plastic or reflective mulches, can deter insects by altering the soil temperature or confusing flying pests.

11. Trapping Heat/Cold

- In buildings, creating extreme temperature zones (hot or cold) can eliminate pests like termites or stored-product pests.
- **Example:** Fumigation chambers that use heat or cold to kill pests in storage facilities.

12. Ultrasound/Noise Repellents

- Devices that emit sound waves or vibrations to repel pests like rodents, birds, or insects.
- Effectiveness varies, and these are typically used as a supplementary measure.

D. Mechanical methods of pest management

Mechanical methods of pest management involve using tools, machinery, or equipment to physically remove, kill, or disrupt pest populations. These techniques are practical, environmentally friendly, and are often part of an integrated pest management (IPM) strategy. Mechanical methods are ideal for controlling pests in small areas, greenhouses, or for specific types of pests. Here are common mechanical methods used in pest management:

1. Handpicking

- **Manual removal** of pests such as insects (e.g., caterpillars, beetles, slugs) from plants.
- **Suitable for small**-scale gardens or low pest populations.
- **Example:** Picking tomato hornworms off plants by hand.

2. Tillage and Cultivation

- **Tillage** involves turning the soil over with plows or tillers to disrupt pest life cycles by exposing them to predators, desiccation, or destruction.
- **Effective against soil**-borne pests, insect larvae, and weeds.
- Cultivating between crop rows disrupts weeds and reduces pest habitats.

3. Mowing and Cutting

- **Mowing** down weeds or vegetation helps reduce habitat for pests, particularly insects and rodents.
- Helps manage pest populations that thrive in tall grasses or weedy areas.

4. Trapping

- **Mechanical traps** are used to capture or kill pests.
  - **Snap Traps**: Used to control rodents like mice or rats.
  - **Pitfall Traps**: Used for crawling insects like beetles.
  - **Insect Traps**: Devices like funnel traps to capture specific insect pests.
- Pheromone traps or baited traps can attract pests to mechanical traps, making them more effective.

5. Barriers and Screens

- **Physical barriers** prevent pests from accessing crops or entering structures.
  - **Fencing**: Protects crops from larger pests like rabbits, deer, or wild boars.
  - **Row Covers**: Fabric covers that prevent insects (e.g., flea beetles, cabbage moths) from laying eggs on crops.
  - **Screens or Nets**: Prevent entry of flying insects like mosquitoes, flies, or moths into greenhouses or storage areas.

6. Vacuuming

- Mechanical **vacuum devices** are used to physically remove pests from plants or structures.
- Commonly used in greenhouses and for crops like strawberries or vegetables to remove pests like aphids, mites, or whiteflies.
- **Example:** "Bug vacuums" used to suck up pests from plants without damaging them.

7. Flaming

- **Flame weeders** use heat from propane burners to kill weeds or pests by direct contact.
- Effective for managing pests like weed seeds and insect eggs on the soil surface.
- Commonly used in organic farming to avoid chemical herbicides.

8. Water Sprays

- **High-pressure water sprays** are used to knock off pests like aphids, mites, and spider mites from plant leaves.

- Safe and non-toxic, this method can be used in gardens and greenhouses.
- **Example:** Washing pests off leaves of crops like tomatoes or roses.

9. Mulching

- **Organic or synthetic mulches** can be placed on the soil surface to create a barrier against pests.
- **Mulches prevent soil**-borne pests, such as certain insect larvae, from reaching crops and suppress weed growth that can harbour pests.

10. Row Covers

- **Floating row covers** are light, breathable fabrics that create a physical barrier between crops and insects or small animals.
- Used to prevent flying insects (e.g., cabbage moths, flea beetles) from laying eggs on crops.

11. Pruning and Thinning

- Removing infected or damaged parts of plants reduces the likelihood of pest infestation spreading.
- Thinning crops (e.g., in fruit orchards) allows better air circulation, which can reduce the prevalence of certain pests like fungal pathogens or insects that thrive in dense foliage.

12. Use of Adhesive Surfaces

- **Sticky traps or tapes** coated with a sticky substance can capture flying pests like whiteflies, aphids, or fungus gnats.
- These are commonly used in greenhouses or indoor spaces where chemicals are less desirable.

13. Screening and Sealing

- Sealing entry points into buildings or greenhouses to prevent pests from entering.
- **Screens on windows** or ventilation systems help keep out flying insects.

14. Soil Barriers

- Use of soil barriers (e.g., sand or gravel) around plant beds or greenhouses to prevent crawling pests like slugs, snails, or ants from reaching crops.

15. Trapping Heat or Cold

- **Cold storage** or heat treatments can be applied in controlled environments (e.g., warehouses) to kill pests in stored products like grains, seeds, or other food products.
- **Example:** Freezing or heating pest-infested products to lethal temperatures.

16. Mechanical Harvesting or Shaking

- Mechanical **shakers** can be used on trees or bushes to dislodge pests (e.g., weevils or mites) from fruit trees.
- These pests can then be collected on tarps or fallen leaves and destroyed.

E. Use of botanicals in pest management

The use of **botanicals** in pest management refers to the application of plant-derived substances to control or repel pests. Botanical pesticides, also known as biopesticides, are an important part of **integrated pest management (IPM)** strategies and offer a natural, environmentally friendly alternative to synthetic chemicals. These substances are generally biodegradable, have low toxicity to non-target organisms, and pose less risk of resistance development in pest populations.

Key Botanicals Used in Pest Management

1. Neem (Azadirachta indica)

- **Active Ingredient**: Azadirachtin.
- **Mode of Action**: Neem has insecticidal, repellent, and antifeedant properties. It disrupts the growth and reproduction of insects, causing larvae to die before becoming adults.
- **Uses**: Effective against a variety of pests such as aphids, whiteflies, leaf miners, and caterpillars. It is also used for soil-borne pests like nematodes.
- **Advantages**: Neem-based products are safe for humans, animals, and beneficial insects like bees.

2. Pyrethrum (Chrysanthemum cinerariifolium)

- **Active Ingredient**: Pyrethrins.
- **Mode of Action**: Pyrethrins attack the nervous system of insects, causing paralysis and death. They are quick-acting and effective on contact.

- **Uses**: Broad-spectrum pesticide used to control pests like mosquitoes, flies, beetles, and caterpillars.
- **Advantages**: Biodegradable and less persistent in the environment than synthetic alternatives. Pyrethrins are often combined with other agents like piperonyl butoxide (a synergist) to enhance effectiveness.
- **Drawbacks**: Toxic to fish and can harm beneficial insects if not applied carefully.

3. Rotenone (Derris elliptica, Lonchocarpus spp.)

- **Active Ingredient**: Rotenone.
- **Mode of Action**: Affects cellular respiration in insects, leading to paralysis and death.
- **Uses**: Effective against a variety of insect pests, including beetles, caterpillars, and aphids. Often used in organic farming.
- **Advantages**: Biodegradable and breaks down quickly in sunlight.
- **Drawbacks**: Rotenone is moderately toxic to humans and animals and has been associated with environmental risks, so its use has declined in recent years.

4. Essential Oils (Eucalyptus, Lavender, Clove, Peppermint, etc.)

- **Active Ingredients**: Various compounds like eugenol (clove oil), limonene (citrus), menthol (peppermint).
- **Mode of Action**: Many essential oils have insecticidal, antifungal, and repellent properties. They can interfere with insect neurotransmitters or disrupt cell membranes.
- **Uses**: Effective against a range of pests, including mosquitoes, ants, and mites. Clove oil, for example, is used to control termites, while peppermint oil is used against ants and aphids.
- **Advantages**: Safe for humans and animals, biodegradable, and often pleasant smelling.
- **Drawbacks**: Essential oils often degrade quickly and may require frequent reapplication for sustained efficacy.

5. Garlic (Allium sativum)

- **Active Ingredient**: Sulfur compounds (like allicin).
- **Mode of Action**: Garlic acts as a repellent by emitting sulfurous compounds that deter insects from feeding or laying eggs.
- **Uses**: Commonly used to repel pests such as aphids, spider mites, and beetles. Garlic sprays can also prevent fungal infections on plants.

- **Advantages**: Non-toxic, environmentally friendly, and safe for beneficial insects.
- **Drawbacks**: Garlic sprays need to be reapplied regularly, particularly after rain.

6. Pepper Extracts (Capsicum spp.)

- **Active Ingredient**: Capsaicin.
- **Mode of Action**: Capsaicin acts as a deterrent by irritating insects and deterring them from feeding on plants.
- **Uses**: Used to manage pests like caterpillars, aphids, and certain beetles. Capsaicin sprays are also used to repel mammals like rabbits and deer.
- **Advantages**: Non-toxic to humans, animals, and the environment.
- **Drawbacks**: Requires frequent application and may be less effective in wet conditions.

7. Sabadilla (Schoenocaulon officinale)

- **Active Ingredient**: Alkaloids like veratrine.
- **Mode of Action**: Sabadilla alkaloids disrupt the nervous system of insects, leading to paralysis and death.
- **Uses**: Effective against pests such as thrips, aphids, leafhoppers, and caterpillars.
- **Advantages**: Breaks down rapidly in the environment and is non-toxic to mammals.
- **Drawbacks**: Can be toxic to beneficial insects and may need frequent reapplication due to rapid degradation.

8. Quassia (Quassia amara)

- **Active Ingredient**: Quassin.
- **Mode of Action**: Quassia acts as an insecticide and repellent, affecting the feeding behavior of insects.
- **Uses**: Used to control pests like aphids, caterpillars, and sawflies.
- **Advantages**: Non-toxic to humans and animals, environmentally safe.
- **Drawbacks**: Requires frequent reapplication.

9. Nicotine (Nicotiana tabacum)

- **Active Ingredient**: Nicotine alkaloids.
- **Mode of Action**: Nicotine affects the nervous system of insects by

blocking acetylcholine receptors, causing paralysis and death.

- **Uses**: Historically used against aphids, thrips, and whiteflies.
- **Drawbacks**: Highly toxic to humans and animals, which has led to its decline in agricultural use.

F. Use of insect pheromones in pest management

Insect pheromones play a key role in pest management as they are highly specific and environmentally friendly tools used to monitor, trap, and control insect pests. Pheromones are natural chemicals emitted by insects to communicate with each other, particularly for mating, aggregation, or marking trails. In pest management, synthetic versions of these pheromones are used in integrated pest management (IPM) programs to reduce or suppress pest populations while minimizing the need for chemical pesticides.

Types of Pheromones used in Pest Management

1. Sex Pheromones

- These are the most commonly used pheromones in pest control and are typically released by female insects to attract males for mating.
- **Mode of Action**: By mimicking the female pheromone, synthetic sex pheromones lure male insects into traps or confuse them, preventing successful mating (known as *mating disruption*).
- **Examples**
  - **Codling Moth (Cydia pomonella)**: Synthetic pheromones are used in orchards to confuse males and reduce mating, effectively controlling codling moth populations in apple and pear orchards.
  - **Gypsy Moth (Lymantria dispar)**: Pheromone traps are used to monitor and suppress populations by disrupting mating behaviour.

2. Aggregation Pheromones

- These pheromones are released to attract both male and female insects to a location, often to food sources or mating sites.
- **Mode of Action**: In pest management, aggregation pheromones are used to attract large numbers of pests into traps, significantly reducing pest populations.
- **Examples**
  - **Bark Beetles (Ips spp.)**: Aggregation pheromones are used to attract beetles into traps, helping to manage infestations in forests.

- **Red Flour Beetle (Tribolium castaneum)**: Used in food storage facilities to control beetle infestations by trapping them.

3. Trail Pheromones

- Used by social insects like ants and termites to mark pathways to food sources.
- **Mode of Action**: In pest management, trail pheromones can be used to lead insects into bait stations or traps. Alternatively, synthetic pheromones can be used to disrupt the trails, leading to disorganization in insect colonies.
- **Examples**
  - **Argentine Ants (Linepithema humile)**: Synthetic trail pheromones can be used to lure ants to bait or disrupt foraging trails.

4. Alarm Pheromones

- Released when insects are disturbed or under threat, warning others in the colony.
- **Mode of Action**: These pheromones are less commonly used in direct pest control but can be studied for potential use in driving pests away or preventing infestations.

Methods of Using Pheromones in Pest Management

1. Monitoring Pests

- Pheromone traps are often used to monitor pest populations. These traps use small amounts of synthetic pheromones to lure pests into a sticky or funnel trap.
- **Purpose**: Monitoring helps farmers and pest control professionals determine when pest populations reach a threshold that requires intervention. It also helps in timing pesticide applications more effectively, reducing unnecessary chemical use.
- **Examples**:
  - Monitoring pests like fruit flies, moths, and beetles in agricultural fields and orchards.
  - **Example**: Pheromone traps are used to monitor Mediterranean fruit flies (Ceratitis capitata) in fruit orchards.

2. Mating Disruption

- **Mating disruption** is a technique where synthetic pheromones are released in large quantities across an area, overwhelming the male insects' ability to locate females for mating. This reduces the overall number of fertilized eggs and, subsequently, the population of the next generation.
- **Examples**
  - **Pink Bollworm (Pectinophora gossypiella)**: Pheromone-based mating disruption has been used successfully in cotton fields to reduce infestations.
  - **Grape Vine Moths (Lobesia botrana)**: Mating disruption techniques are used in vineyards to protect grape crops.

3. Mass Trapping

- **Mass trapping** uses large numbers of pheromone traps to attract and catch a significant portion of the pest population. This method can be used for population control or eradication efforts.
- **Examples**
  - **Mediterranean Fruit Fly**: Pheromone-baited traps are used in combination with other control measures to reduce fruit fly populations in citrus orchards.
  - **Palm Weevils (Rhynchophorus ferrugineus)**: Pheromone traps are used to control this pest in palm trees, particularly in date palm plantations.

4. Lure-and-Kill

- This method combines pheromones with a small dose of insecticide. Pests are attracted by the pheromones to a bait station, where they come into contact with the insecticide and die.
- **Examples**
  - **Tsetse Fly Control**: Pheromone baits combined with insecticides have been used in Africa to control tsetse flies, which spread diseases like sleeping sickness.

Advantages of Using Pheromones in Pest Management

1. Environmentally Friendly

- Pheromones are species-specific and target only the pest species of concern, reducing the impact on beneficial insects, pollinators, and other non-target organisms.

- Pheromones break down naturally in the environment, leaving no harmful residues.

2. Reduced Pesticide Use

- By using pheromones to monitor or control pest populations, farmers can reduce or eliminate the need for chemical insecticides. This leads to lower pesticide residues on crops and in the environment.

3. Low Risk of Resistance

- Since pheromones are natural chemicals used in low quantities, pests are less likely to develop resistance to pheromone-based control methods, unlike with conventional chemical pesticides.

4. Safe for Humans and Animals

- Pheromones are non-toxic to humans, animals, and other wildlife. They are safe for use in residential, agricultural, and public spaces.

5. Effective Population Management

- Pheromone-based techniques like mass trapping or mating disruption can be highly effective in reducing pest populations over time, helping to protect crops and reduce economic losses.

Challenges of Using Pheromones in Pest Management

1. Cost

- The production and deployment of synthetic pheromones can be expensive, especially for large-scale applications. Mating disruption and mass trapping often require extensive equipment and regular maintenance.

2. Labor-Intensive

- Some pheromone traps or mating disruption devices require frequent monitoring, replacement, and proper placement to be effective.

3. Limited Range

- Pheromones typically have a limited effective range, which means they may need to be deployed in large numbers to cover an entire field or orchard effectively.

4. Species-Specific

- Pheromones are highly specific to particular species, which is an advantage but can also be a limitation if multiple pest species need to be managed simultaneously.

Examples of Successful Pheromone Use

- **Codling Moth Control in Apple Orchards**: Pheromone traps and mating disruption have been successfully used to control codling moth populations in apple orchards, reducing the need for pesticide sprays.
- **Mating Disruption for Grape Vine Moths**: Vineyards use pheromones to control grape vine moths, significantly lowering damage to grape crops and reducing insecticide use.
- **Mass Trapping of Red Palm Weevil**: In date palm plantations, pheromone traps have helped control the spread of the destructive red palm weevil.

G. **Biological control of pest**

Biological control is a method of pest management that uses living organisms—such as predators, parasites, pathogens, or competitors—to control pest populations. It is an environmentally sustainable alternative to chemical pesticides and a key component of **Integrated Pest Management (IPM)** programs. Biological control can help maintain ecological balance, reduce pesticide use, and minimize harm to non-target species, including humans, animals, and beneficial organisms.

Types of Biological Control

1. Classical Biological Control

- Involves introducing a natural enemy (predator, parasitoid, or pathogen) from a pest's native range into a new area where the pest has become invasive.
- This method is typically used for long-term control of non-native or invasive pests that lack natural enemies in the new environment.
- **Example**: Introduction of the *Vedalia beetle* (Rodolia cardinalis) from Australia to California to control cottony cushion scale (Icerya purchasi), a pest on citrus crops.

2. Augmentative Biological Control

- Involves the **supplemental release** of natural enemies to boost existing populations or establish new populations for short-term control of pests.

- There are two subtypes:
  - **Inoculative Releases**: Small numbers of natural enemies are released early in the pest infestation to build up populations over time.
  - **Inundative Releases**: Large numbers of natural enemies are released to quickly reduce pest populations.
- **Example**: Releasing *Trichogramma* wasps to parasitize the eggs of caterpillar pests like European corn borer.

3. Conservation Biological Control

- Focuses on **enhancing the effectiveness of natural enemies** already present in the environment by conserving or promoting their populations.
- Strategies include reducing pesticide use, providing habitat (such as hedgerows or cover crops), and ensuring alternative food sources.
- **Example**: Conserving ladybugs (lady beetles) and lacewings in agricultural fields to control aphid populations.

Types of Biological Control Agents

1. Predators

- Predators are free-living organisms that feed on a variety of prey (pests) throughout their lifetime. They often kill and consume multiple pest individuals.
- **Examples**
  - **Ladybugs (Coccinellidae)**: Predators of aphids, mealybugs, and other soft-bodied insects.
  - **Lacewings (Chrysopidae)**: Larvae of lacewings are voracious feeders of aphids, mites, and small caterpillars.
  - **Spiders**: Generalist predators that feed on various insect pests.
  - **Predatory Beetles**: Ground beetles (Carabidae) and rove beetles (Staphylinidae) prey on insect larvae, slugs, and snails in soil.

2. Parasitoids

- Parasitoids are organisms that lay their eggs inside or on a host pest. The developing larvae feed on and eventually kill the host.
- Parasitoids are usually specific to a particular pest or a group of closely related pests.

- **Examples**:
  - **Trichogramma Wasps**: Parasitoids of moth and butterfly eggs, commonly used to control caterpillar pests like the corn borer and cabbage looper.
  - **Braconid and Ichneumonid Wasps**: Parasitoid wasps that target aphids, caterpillars, and beetles.
  - **Tachinid Flies**: Parasitoids of caterpillars, beetle larvae, and other insect pests.

3. Pathogens

- Pathogens are microorganisms such as bacteria, viruses, fungi, and protozoa that cause diseases in pest insects. Once the pest is infected, the pathogen multiplies and eventually kills it.
- **Examples**
  - **Bacillus thuringiensis (Bt)**: A bacterial pathogen that produces toxins harmful to caterpillars, beetles, and mosquito larvae. Bt is widely used as a microbial insecticide in organic farming.
  - **Entomopathogenic Fungi**: Fungi such as *Beauveria bassiana* and *Metarhizium anisopliae* infect and kill a wide range of insects, including aphids, whiteflies, and termites.
  - **Nuclear Polyhedrosis Virus (NPV)**: A viral pathogen used to control caterpillar pests like the gypsy moth and armyworm.

4. Nematodes

- Entomopathogenic nematodes (roundworms) are soil-dwelling organisms that parasitize insect larvae, especially those that live in the soil.
- **Examples**
  - **Steinernema and Heterorhabditis** species: These nematodes infect and kill the larvae of insects like rootworms, grubs, and cutworms by releasing symbiotic bacteria that help kill the host.
  - Nematodes are often used in lawns, turf, and greenhouses to control soil-borne pests.

5. Competitors

- Competitors are species that outcompete pests for food, habitat, or other resources, indirectly reducing pest populations.

- **Example**: In the control of weeds (which are considered pests in agriculture), certain cover crops or plants can outcompete and suppress weed growth.

Advantages of Biological Control

1. Environmentally Safe

- Biological control agents are natural enemies of pests and pose minimal risk to humans, animals, and beneficial organisms. They are typically species-specific and have a low environmental impact.

2. Sustainable

- Once established, biological control agents can often provide long-term pest suppression with minimal additional inputs, reducing reliance on chemical pesticides.

3. No Resistance Development

- Unlike chemical pesticides, which pests can develop resistance to over time, biological control methods rarely lead to resistance in pest populations because they involve natural ecological processes.

4. Preserves Ecosystem Health

- By maintaining biodiversity and reducing the use of broad-spectrum pesticides, biological control helps preserve beneficial insects, pollinators, and soil health.

5. Cost-Effective

- Although the initial setup cost can be high (especially in classical biological control), once established, natural enemies can self-replicate and provide ongoing control without the need for repeated applications like chemical pesticides.

Challenges of Biological Control

1. Slower Action

- Biological control methods often take longer to achieve noticeable results compared to chemical pesticides. Natural enemies need time to establish and spread within the pest population.

2. Requires Specific Conditions:

- The effectiveness of biological control can depend on environmental factors like temperature, humidity, and the availability of suitable habitats for natural enemies.

3. Limited to Targeted Pests

- Biological control agents are often specific to certain pests, meaning they might not be suitable for controlling multiple pest species simultaneously.

4. Potential Non-Target Effects

- In some cases, introduced biological control agents may affect non-target species or become invasive themselves if not properly researched and regulated. However, modern biological control programs are highly regulated to minimize such risks.

5. Seasonality and Timing

- The activity and effectiveness of natural enemies can be influenced by seasonal factors. In some cases, augmentative releases are necessary to synchronize with the pest's life cycle.

Examples of Successful Biological Control Programs

1. Control of Cottony Cushion Scale

- The introduction of *Rodolia cardinalis* (a ladybird beetle) into California citrus orchards in the late 19th century was one of the earliest and most successful examples of classical biological control. The beetle effectively controlled cottony cushion scale, saving the citrus industry from severe damage.

2. Control of Cassava Mealybug

- In Africa, the release of the parasitoid wasp *Anagyrus lopezi* successfully controlled the invasive cassava mealybug (Phenacoccus manihoti), which had been causing widespread damage to cassava crops.

3. Use of Trichogramma Wasps

- *Trichogramma* wasps are commonly used to control the eggs of lepidopteran pests (moths and butterflies) in crops like maize, cotton, and vineyards. The wasps parasitize the eggs of pests such as the European corn borer and the cabbage looper.

H. Indigenous Technical Knowledge

The knowledge of traditional agriculture with millions of farmers should be utilized and modern technology in agriculture should be blended with traditional wisdom. The following are certain practices of farmers which they have been following time immemorial

- Diluted cow dung slurry sprinkled to hasten paddy germination. • Coconut fronds cut into small bits erected as perches in field to attract nocturnal birds which preys upon rats. • Chilli mash and garlic juice sprayed to control rice earhead bug.
- Application of common salt at 1 - 1.5 kg/ palm of coconut gives insect resistance and prevents button shedding.
- Use of scarecrows to ward off bird pests in day time, which also serve as perches to nocturnal predatory birds.
- Use of Kavankal where stones are released from slings to scare away birds
- Ploughing of field during summer months (April-May) (Agninakshatra) when temperature is around 40 - 45 °C brings about killing of soil insects, pathogens, nematodes and pupae of lepidopteran pests.
- Treating stored pulses with red earth to prevent insect damage.
- Use of Tanjore bow trap, a common traditional gadget to kill rats in rice fields of Cauvery delta.
- 'Vrikshayurveda", a science of plant health, similar to 'Ayurveda" which is science of human life deals with maintenance of plant health and provides literature on control measures for control of pests and diseases. There are many such practices based on traditional wisdom of farmers in different regions of the country. Scientific bases behind such practices if established based on research would help in including them in management measures.

7

Role of Agroforestry in Natural Farming

Agroforestry plays a vital role in natural farming by integrating trees and shrubs with crops and livestock in ways that create diverse, productive, and sustainable land-use systems. In natural farming, where minimal chemical inputs are used, agroforestry offers numerous environmental and economic benefits, making it a valuable approach for sustainable agriculture.

1. Soil Health and Fertility

Soil health and fertility are the cornerstones of any sustainable agricultural system, particularly in natural farming, where synthetic inputs are minimized or avoided altogether. Agroforestry, an agricultural practice that integrates trees and shrubs into farming systems, plays a key role in maintaining and enhancing soil health and fertility through natural processes. Here's how agroforestry significantly improves soil conditions and sustains fertility over time. One of the primary ways agroforestry enhances soil fertility is through nutrient cycling. Trees and shrubs shed leaves, twigs, and other organic matter onto the soil surface, forming a layer of mulch. As this organic matter decomposes, it releases essential nutrients such as nitrogen, phosphorus, and potassium back into the soil, which are then available for uptake by crops. This process mimics the natural nutrient recycling seen in forests, reducing the need for chemical fertilizers and supporting the goal of low-input natural farming. Some tree species used in agroforestry are nitrogen-fixing, such as certain types of legumes like Gliricidia and Leucaena. These trees form symbiotic relationships with nitrogen-fixing bacteria in their root nodules, converting atmospheric nitrogen into forms that plants can use. As a result, these trees contribute nitrogen directly to the soil, boosting its fertility and helping to meet the nitrogen needs of companion crops without synthetic fertilizers. Agroforestry improves soil structure by encouraging the formation of stable soil aggregates. Tree roots penetrate deep into the soil, creating channels that enhance aeration and water infiltration. This process not only reduces surface runoff and soil erosion but also supports the growth of beneficial soil organisms, such as earthworms and mycorrhizal fungi, which further improve soil structure. Better soil structure also helps crops develop strong root systems, which are essential for accessing water and nutrients deep within the soil. In regions prone to erosion, tree roots anchor the soil, preventing it from being washed or blown away. Trees act as windbreaks, protecting crops from wind erosion, and their canopies intercept rainfall, lessening the impact of raindrops on the soil surface. By reducing both water and wind erosion, agroforestry systems preserve topsoil, which contains the highest concentration of nutrients and organic matter crucial for crop growth. The presence of trees in agroforestry systems can significantly improve the soil's capacity to retain water. Tree canopies moderate soil temperature by providing shade, reducing evaporation, and maintaining higher soil moisture levels. Additionally, organic matter from decomposing plant materials increases the soil's water-holding capacity, making it more resilient to drought. This enhanced water retention supports crop productivity, especially in rain-fed systems, and aligns with the goals of natural farming to reduce dependency on external water sources. Healthy soil biodiversity is essential for sustaining soil fertility over time. Agroforestry systems create a hospitable environment for a diverse range of soil organisms, including bacteria, fungi, insects, and earthworms. These organisms play a crucial role in decomposing organic matter, recycling nutrients, and promoting plant health. For example, mycorrhizal fungi form beneficial relationships with plant roots, aiding in nutrient absorption, especially phosphorus, and improving plant resilience to stress. By supporting a rich soil food web, agroforestry fosters a self-sustaining ecosystem that maintains soil fertility naturally. Trees in agroforestry systems sequester carbon from the atmosphere, contributing to climate change mitigation. The carbon stored in tree biomass and added to the soil through organic matter decomposition increases soil organic carbon levels, which is a key indicator of soil fertility. Higher soil organic carbon improves soil structure, water retention, and nutrient availability, benefiting crops. In summary, agroforestry enhances soil health and fertility through natural processes that build a robust, self-sustaining soil ecosystem. This approach not only reduces the need for external inputs but also builds resilience against environmental stresses, making agroforestry a valuable practice in natural farming. By supporting nutrient cycling, preventing erosion, retaining water, promoting biodiversity, and sequestering carbon, agroforestry offers a holistic, sustainable method for maintaining fertile, productive soils.

2. Water Conservation

Water conservation is a crucial component of sustainable agriculture, especially in regions prone to water scarcity or drought. In natural farming systems, which emphasize low external input use and ecosystem harmony, agroforestry plays a key role in conserving water through various ecological mechanisms. By integrating trees and shrubs with crops and livestock, agroforestry enhances water retention, reduces evaporation, and improves water availability for plants, contributing to a resilient and sustainable farming system. Trees in agroforestry systems create shade, reducing direct sunlight exposure on the soil surface. This shading effect lowers soil temperatures, which in turn decreases the rate of water evaporation from the soil. In addition to shading, tree canopies intercept rainfall, slowing its impact on the ground and reducing the splash effect that can displace soil particles and moisture. This shade combined with soil moisture retention supports longer-lasting soil moisture levels, especially beneficial during dry periods. Maintaining higher soil moisture allows crops to thrive with less water and aligns with natural farming's objective of minimizing reliance on external irrigation. Tree roots improve soil structure by creating channels that enhance water infiltration. These roots reach deep into the soil, loosening compact layers and allowing rainwater to permeate more effectively. This water percolates down into deeper soil layers where it can be stored and accessed by crops during periods of limited rainfall. Enhanced water infiltration also prevents waterlogging and runoff, which are common in compacted or bare soils. With improved infiltration, agroforestry systems make more effective use of available water, reducing the need for frequent irrigation and preserving soil health. Soil erosion, a major contributor to water loss, is significantly reduced in agroforestry systems. Tree roots anchor soil particles, preventing them from being washed away by rainwater. Tree canopies also act as a buffer for raindrops, lessening their impact on the soil surface. As a result, agroforestry systems experience less surface runoff, allowing water to remain in the soil rather than being lost to surrounding areas. In sloped landscapes or areas prone to erosion, strategically planted trees reduce soil erosion, conserve water, and protect downstream water resources. The deeper rooting systems of many tree species help draw water up from deep underground layers, improving water availability in the soil. Some of this water is eventually released through transpiration, a process that maintains local humidity and can support rain formation. By encouraging water infiltration, agroforestry helps replenish groundwater supplies, which are vital for long-term water security. In areas where rainfall is seasonal, agroforestry systems can support the slow, steady recharge of groundwater, which is particularly valuable for water conservation in drought-prone regions. Agroforestry systems contribute organic matter to the soil in the form of leaf litter and decomposed plant materials. This organic matter, when left on the soil surface, acts as mulch, covering the soil and protecting it from direct sunlight. Mulch helps to retain soil moisture by reducing evaporation and creates a barrier that prevents soil compaction and erosion. As the mulch decomposes, it also improves soil structure, enhancing the soil's ability to retain water. This continuous supply of organic mulch reduces the need for artificial mulching materials and conserves water naturally. Trees and shrubs in agroforestry systems create a microclimate that promotes higher humidity levels, which can reduce water stress for crops. By breaking wind and creating shade, trees moderate extreme temperature fluctuations, reducing plant water loss through transpiration. The microclimate effect is particularly beneficial in arid and semi-arid regions, where high temperatures and dry winds can quickly deplete soil moisture. In summary, agroforestry systems provide a multifaceted approach to water conservation, making them ideal for natural farming. Through reduced evaporation, enhanced infiltration, erosion control, groundwater recharge, organic mulching, and microclimate formation, agroforestry supports the efficient use and conservation of water. These benefits help create resilient farming systems capable of withstanding climate variability and contribute to the long-term sustainability of natural farming practices.

3. Pest and Disease Management

In natural farming, pest and disease management focuses on creating balanced ecosystems that naturally keep pests and diseases in check without relying on synthetic pesticides. Agroforestry, which integrates trees, shrubs, crops, and sometimes livestock into a single system, supports this approach by enhancing biodiversity, fostering beneficial organisms, and disrupting pest lifecycles. Agroforestry promotes biodiversity, introducing a variety of plants that attract diverse organisms. These organisms include predators of common agricultural pests, such as ladybugs, spiders, birds, and parasitoid wasps. By providing habitats and food sources for these beneficial insects and animals, agroforestry systems establish a natural pest control mechanism that minimizes pest populations. For example, birds that nest in agroforestry trees may consume insects that would otherwise damage crops. This predation reduces the need for chemical interventions, aligning with natural farming's low-input approach. In monoculture systems, pests can easily locate and attack their host crops, leading to higher pest pressure. Agroforestry disrupts these cycles by intermixing different plant species, making it harder for pests to locate their target plants. This mixed planting approach, often referred to as "spatial heterogeneity, " limits the spread of pest populations by breaking up their habitat. For example, intercropping legumes with fruit trees or vegetables can break the cycle of pests that specialize in one crop, as they are forced to travel further between plants. Agroforestry systems that incorporate flowering plants attract pollinators like bees and butterflies, which are crucial for the reproduction of many crops. Some flowering plants, such as clover or sunflowers, also attract beneficial insects that prey on common pests. By providing habitats for both pollinators and predatory insects, agroforestry not only supports healthy crop yields but also builds a balanced ecosystem that discourages pest outbreaks. In agroforestry systems, the physical separation between crop plants can also act as a barrier against the spread of plant diseases. Trees and shrubs form natural windbreaks, reducing the likelihood that disease spores or infected insects will spread from one area to another. This separation is particularly useful for controlling airborne diseases and fungal infections, as it minimizes cross-infection among plants. Healthy plants are more resistant to pests and diseases, and agroforestry contributes to this resilience through improved soil health. The organic matter from decomposing leaves and tree litter enhances soil structure and nutrient availability, which strengthens plant health. Well-nourished plants can better withstand pest attacks and diseases, as they have stronger natural defenses. In addition, some trees support beneficial soil organisms, such as mycorrhizal fungi, which aid in nutrient absorption and can help plants resist soil-borne diseases. Some trees and shrubs commonly used in agroforestry, such as neem, moringa, and basil, have pest-repellent properties. These plants naturally deter pests from the area through compounds they release, which repel or disrupt pest behavior. Strategically planting these species within agroforestry systems offers an added layer of pest control, further reducing the need for pesticides.

4. Carbon Sequestration

Carbon sequestration is the process of capturing and storing atmospheric carbon dioxide (CO_2) in plants, soils, oceans, and geological formations. In agriculture, particularly in natural farming systems, carbon sequestration is essential for combating climate change and improving soil health. Agroforestry, the integration of trees and shrubs with crops and livestock, plays a significant role in carbon sequestration by incorporating more biomass into farming systems. This approach not only helps reduce the amount of carbon in the atmosphere but also provides other ecological and agricultural benefits. Trees absorb CO_2 during photosynthesis, storing it as carbon in their trunks, branches, leaves, and roots. This biomass acts as a carbon sink, holding carbon that would otherwise remain in the atmosphere. Agroforestry systems incorporate trees and shrubs into crop fields or pasturelands, providing continuous carbon storage. Over time, as trees grow larger, they store more carbon, creating a long-term carbon reserve. Compared to conventional agriculture, which often lacks perennial vegetation, agroforestry systems have a greater capacity for carbon storage, as trees remain in place for decades. In addition to storing carbon in their biomass, agroforestry systems increase soil organic carbon (SOC) through leaf litter, root decay, and organic matter deposition. When tree leaves, twigs, and roots decompose, they enrich the soil with organic matter. This organic carbon becomes part of the soil, enhancing soil health and fertility while serving as a stable form of carbon storage. Roots play a critical role in building SOC as they penetrate deeper layers of soil. The carbon captured in these roots is often stored for long periods, as deep soil layers are less susceptible to disturbances like erosion or decomposition. Additionally, tree roots improve soil structure, making it more resilient and less prone to carbon loss. By increasing SOC, agroforestry also supports healthier and more productive soils, which is essential for sustainable farming and improving crop resilience. Agroforestry reduces greenhouse gas emissions by providing an alternative to carbon-intensive farming practices. By enhancing soil health, agroforestry reduces the need for synthetic fertilizers, which are often associated with high greenhouse gas emissions during production and application. In natural farming, where minimizing inputs is a priority, agroforestry helps farmers meet crop nutrient needs more sustainably. Agroforestry trees create shade and act as windbreaks, reducing temperature extremes and protecting crops from wind damage. This microclimate can lower water evaporation rates and reduce irrigation needs, which conserves energy. Reducing energy use indirectly decreases emissions from fossil fuel-powered equipment, such as water pumps and tractors, further minimizing the system's overall carbon footprint. By creating diverse habitats, agroforestry supports biodiversity, which enhances the resilience of the ecosystem. Diverse ecosystems are better able to withstand environmental stresses, such as pest outbreaks and extreme weather. This resilience reduces the likelihood of crop loss, which in turn lessens the need for resource-intensive recovery efforts. With a more stable ecosystem, agroforestry systems sequester more carbon and maintain their carbon stores even under changing climate conditions.

5. Biodiversity Conservation

Biodiversity conservation is essential for sustaining ecosystems and agricultural resilience, particularly in natural farming systems that aim to minimize external inputs and foster ecological balance. Agroforestry, the practice of integrating trees, shrubs, crops, and sometimes livestock on the same land, plays a crucial role in biodiversity conservation by providing habitats, promoting species diversity, and enhancing ecosystem stability. Agroforestry systems provide a range of habitats for various organisms. The inclusion of trees, shrubs, and ground cover creates multiple layers of vegetation, each offering shelter, food, and breeding grounds for different species, from insects and birds to mammals and soil microbes. Trees, in particular, support birds, small mammals, and beneficial insects that rely on foliage, flowers, and fruits as food sources. Additionally, some species find nesting and roosting sites in tree canopies, making agroforestry farms ideal refuges for wildlife. Pollinators such as bees, butterflies, and birds play a vital role in agricultural productivity. Agroforestry plants, especially flowering trees and shrubs, provide pollen and nectar sources that support these pollinators throughout their life cycles. By fostering pollinator populations, agroforestry boosts crop yields and strengthens food security. Additionally, agroforestry systems attract natural predators of agricultural pests. Predators like birds, spiders, and ladybugs help control pest populations naturally, reducing the need for chemical pesticides. This biological pest control is a core principle in natural farming, as it prevents pests from becoming resistant to chemical treatments and protects beneficial insects. Agroforestry encourages the growth of multiple crop and tree species, contributing to genetic diversity within the farming system. A diverse genetic pool provides resilience against environmental stressors like drought, pests, and diseases. For instance, if a particular pest or disease affects one crop species, other crops in the system may remain unaffected, safeguarding yields and reducing economic risks for farmers. Soil health is closely linked to biodiversity, particularly in terms of soil organisms like bacteria, fungi, and earthworms. Agroforestry practices contribute organic matter through leaf litter and root decay, enriching the soil and supporting these organisms. Healthier soils promote plant biodiversity and, in turn, a more balanced ecosystem that supports diverse life forms. Biodiversity in agroforestry systems contributes to climate resilience. Diverse ecosystems are more adaptable to climate change impacts like extreme weather, drought, and shifting pest patterns. Agroforestry's multi-species composition can buffer against these changes, offering habitats for species displaced by environmental shifts and maintaining overall ecosystem stability.

6. Economic Resilience

Economic resilience in agriculture refers to the ability of farming systems to withstand and recover from economic shocks and stresses, such as price fluctuations, climate events, and changing market demands. Agroforestry, the integration of trees, shrubs, crops, and sometimes livestock on the same land, enhances economic resilience by diversifying income sources, reducing input costs, and increasing productivity. This approach aligns well with natural farming, where the focus is on reducing external inputs and creating self-sustaining systems. Agroforestry systems are inherently diverse, incorporating multiple crops, trees, and sometimes livestock on the same land. This diversity provides farmers with a range of marketable products, such as fruits, nuts, timber, fuelwood, and fodder, in addition to traditional crops. By producing several goods instead of relying on a single crop, farmers can spread their financial risk. If one crop fails due to disease, pests, or market fluctuations, other products may still provide income, creating a financial safety net. In natural farming, reducing dependence on external inputs like synthetic fertilizers and pesticides is key. Agroforestry contributes to this by enhancing soil fertility, supporting natural pest control, and improving soil moisture retention. Trees in agroforestry systems contribute organic matter, such as leaves and decomposing branches, which builds soil fertility and reduces the need for purchased fertilizers. Some trees, such as nitrogen-fixing legumes, enrich the soil naturally by fixing atmospheric nitrogen, further decreasing fertilizer requirements. Trees in agroforestry systems provide shade, reduce wind speeds, and create a more stable microclimate, protecting crops from extreme weather events like heat waves, strong winds, and drought. For example, shade from trees can prevent excessive evaporation, helping conserve soil moisture and reduce irrigation needs. Tree roots improve soil structure and enhance water infiltration, making the soil more resilient to heavy rains or droughts. By reducing environmental stress, agroforestry systems often lead to more stable yields, even under adverse weather conditions, ensuring a more reliable food and income source for farmers. Trees in agroforestry systems, particularly those used for timber, represent a valuable long-term asset. As these trees mature, they increase in value, providing a form of "savings" that can be accessed during difficult times. For instance, farmers can selectively harvest timber to generate income during an economic downturn, cover unexpected expenses, or invest in farm improvements. This long-term asset-building aspect of agroforestry helps farmers strengthen their economic resilience by giving them resources to rely on during periods of economic hardship. With a diversified product range, agroforestry farmers can tap into various markets, from food and fiber to sustainable products like organic and sustainably grown goods. Some agroforestry products, such as non-timber forest products or specialty crops like medicinal plants, attract niche markets that offer premium prices. This access to different markets enhances farmers' adaptability to economic shifts and allows them to reach higher-value customers who are interested in sustainably produced goods.

7. Microclimate Moderation

Microclimate moderation is the process of creating a stable, localized climate within a specific area, which can protect crops from extreme temperature fluctuations, winds, and evaporation rates. Agroforestry, a farming system that integrates trees and shrubs with crops and sometimes livestock, plays a significant role in moderating microclimates. By enhancing temperature stability, improving humidity, and reducing wind exposure, agroforestry helps create a favorable environment for crops and animals. This practice is particularly valuable in natural farming systems, where farmers aim to reduce reliance on external inputs and improve resilience to climate stress. Trees in agroforestry systems provide shade, which helps to lower soil and air temperatures during hot periods. The shade prevents the sun from directly hitting the soil and crops, reducing the risk of heat stress, sunburn, and moisture loss in plants. By cooling the soil surface, trees also reduce soil temperature fluctuations, which helps maintain soil structure and protects root systems from extreme temperatures. This temperature stability is particularly beneficial in hot climates, where high temperatures can damage crops, reduce growth rates, and even lead to crop failure. During cooler months, trees act as a buffer against cold winds and frost, protecting crops from chilling injuries. In this way, agroforestry helps extend the growing season and improves the consistency of crop yields, making farming more resilient to seasonal weather fluctuations. Trees in agroforestry systems contribute to higher humidity levels by transpiring water through their leaves. Transpiration releases moisture into the air, increasing relative humidity around crops and reducing plant water stress. This added moisture also creates a damp environment that lowers evaporation rates from the soil, conserving valuable water resources. When crops are protected from excessive evaporation, they require less irrigation, which is beneficial for natural farming systems that aim to minimize water usage. Additionally, leaf litter and organic mulch from trees help retain soil moisture. When fallen leaves decompose, they create a layer of organic matter that reduces water evaporation and keeps the soil damp. Microclimate moderation through agroforestry is particularly important in the face of climate change. As global temperatures rise and weather patterns become more unpredictable, the ability of agroforestry systems to buffer extreme conditions becomes increasingly valuable. By creating favorable microclimates, agroforestry can help mitigate the impacts of climate variability, allowing farmers to adapt their practices and sustain productivity despite changing environmental conditions. Trees release water vapor into the atmosphere through a process called transpiration. This process increases local humidity levels, creating a more favorable microclimate for plants. Higher humidity can reduce plant stress and improve photosynthesis, which is essential for crop growth. In agroforestry systems, the increased humidity provided by trees can help crops thrive, particularly in areas where low humidity is detrimental to plant health. This regulation of humidity contributes to a more balanced ecosystem, benefiting both crops and livestock.

8

Disease Management in Natural Farming

When a plant is unable to perform its functions normally, then it is diseased. The two major causes of disease in crops and trees are pathogens and environmental conditions. There are many kinds of disease pathogens: viruses, bacteria, fungi, nematodes, mycoplasma-like organisms, and parasitic higher plants. Fungal pathogens are the most common type. They cause seed rots, seedling damping-off, root rots, foliage diseases, cankers, vascular wilts, diebacks, galls and tumors, trunk rots, and decay of aging trees. Deterrents to establishment include unfavourable weather and environmental conditions such as Temperature and moisture extremes, high winds and ice may directly injure a tree and predispose the trees to pest attack.

Biodiversity plays multiple roles in development as well as ecological sustainability. It has received increased attention in agricultural ecosystems for the past few years regarding the reduction of plant diseases; however, the knowledge of using biodiversity to manage plant diseases is Still incomplete. For crop diversification, the interaction of genetic diversity per se in host populations with identifiable resistance and other functional traits of component genotypes to mitigate disease epidemics is not well understood, and neither is the best way of structuring mixed populations. The mixing of crop varieties can significantly reduce disease epidemics in the field. To achieve the best disease mitigation, growers should include as many varieties as possible in mixtures or, if only two component mixtures are possible. The many strategies, tactics, and techniques used in disease management can be grouped under very broad principles of action. The differences between these principles are often not clear. The simplest system consists of two principles, prevention (prophylaxis in some early writings) and therapy (treatment or cure). The first principle (prevention) has to do with disease management practices applied before infection-that is, the plant is protected from disease. The second principle (therapy or curative action) works with any measure applied after the plant is infected-that is, the plant is treated for the disease. For example, enforcement of quarantines prevents introduction of a disease agent (pathogen) into a region where it does not occur. The second principle is represented by heat or chemical treatment of vegetative material like bulbs, corms, and woody cuttings to kill fungi, bacteria, nematodes, or viruses that are already established in the plant material. In another method, there are four general disease control principles. These are: exclusion; eradication; protection, and immunization-the last principle is better called resistance since plants lack an immune system in the same way that animals do. Such principles are elaborated herein.

Exclusion : This principle is defined as any measure that prevents the entry of a disease-causing agent, or pathogen, into a region, farm, or planting. The basic strategy assumes that most pathogens can only travel short distances without some other agent to assist such as humans or other vector. Many pathogens travel along with their host plants, even on material such as soil, packing material or containers. Alas, exclusion measures typically will only delay the entry of a pathogen, but exclusion may provide time to plan how to manage the pathogen when it arrives finally. An important and practical strategy for excluding pathogens is to produce pathogen-free seed or planting stock through certification programs for seeds and vegetatively Plant material, such as potatoes, grapes, tree fruits, etc. These programmes exploit the use of application of technologies including isolation of production areas, field inspections and eradication of suspect plants to produce and maintain pathogen-free stocks. Exclusion can be as simple as cleaning farming equipment to eliminate contaminated debris and soil that may harbour pathogens such as Verticillium, nematodes or other soilborne organisms. This would thereby prevent them from entering infestation-free fields.

Eradication : This principle is a try to eliminate a pathogen after it has been introduced to an area but before becoming established or highly disseminated. It can be applied at the individual plant, seed lot, field or regional level, but generally is not effective over large geographic areas. Eradication of citrus canker involved significant removal and burning of infected trees and, in some instances, whole citrus groves and nurseries. The disease is transmitted, seemingly eradicated, by the pathogen, but the disease resurfaces and efforts to eradicate are continuing.

Elimination might also be on a more modest scale like eradication of apple or pear. Or perhaps the removal of branches infected by the fire blight bacterium, Erwinia amylovora or pruning of blister rust cankers caused by Cronartium ribicola on white pine branches. Perhaps it is just sorting out and removal of diseased flower bulbs, corms, or rhizomes. Seed treatment of cereals to kill mycelium in the seeds and disinfecting with heat to remove the viruses from Some examples of pathogen eradication include fruit tree budwood for grafting. Another mode of eradication can be done through the removal of weeds that harbor the reservoir of several pathogens or vectors. Removal of cull piles of potato is one efficient means of eradicating the surviving inoculum of the late blight pathogen from overwintering. Crop rotation is perhaps the most commonly employed method to reduce the availability of pathogens. Normally soilborne organisms within a cropping area. Examples include take-all of wheat caused by Gaeumannomyces graminis and soybean cyst nematode by Heterodera glycines that are easily controlled by rotations of 1 and 2 years respectively out of the susceptible crops, which include susceptible weed hosts like grasses for take-all.

Protection : This is based on the principle of segregation of pathogen from host. A plant or the susceptible portion of the host plant. Besides chemical defenses, physical Thus, in addition to barriers, explicit techniques taken to assume that pathogens are That infection will manifest and occur without any protective factors at play. For for instance, bananas receive plastic sleeves once fruits are set to cover Fruits from other pests, including fruit decay fungi. Protections typically encompass some type of cultural activity that modifies the environment, such as as tillage, drainage, irrigation, or altering soil pH. It can further have date or depth.

For instance, seeding and spacing of the plants, pruning and pruning, and the rest-whatever technique might help the plants avoid To cause infection or weakening of the disease. Raise planting beds in order to improve water status of soil drainage is a good example of cultural management of plant diseases such as root and stem rots.

Resistance : Disease-resistant plants are the best control measure for diseases in plants, if plants

That is of suitable quality for the growing area, with adequate levels of persistent resistance is available. The use of disease-resistant plants eliminates the need for additional efforts to reduce disease losses unless other diseases are additionally present. Resistant plants such are generally acquired by conventionally bred practices involving selection and/or hybridization.

Integrated Disease Management

Integrated Disease Management (IDM) consists of scouting with timely application of a combination of strategies and tactics. These may include site selection and preparation, utilizing resistant cultivars, altering planting practices, modifying the environment by drainage, irrigation, pruning, thinning, shading, etc. However, in NF chemicals are not applied. But in addition to these traditional measures, monitoring environmental factors (temperature, moisture, soil pH, nutrients, etc.), disease forecasting, and establishing economic thresholds are important to the management scheme. These measures should be applied in a coordinated integrated and harmonized manner to maximize the benefits of each component. Cultivating healthy vigorous plants reduces the disease incidence. However, this is not always easy to accomplish, and "disease management" may be reduced to single measures exactly the same as the ones previously called "disease control." Whatever the measures used, they must be compatible with the cultural practices essential for the crop being managed. Traditionally, there are several types of crop diseases: abiotic (also known as noninfectious) and biotic (infectious). Unfavourable environmental conditions often generate non-communicable diseases. Examples are low or high temperature, excess, or lack of moisture. Also, diseases are usually caused by harmful impurities in the air. They can accumulate due to the presence of nearby chemical or metallurgical plants. Usually, the unhealthy physicochemical composition of the soil is the disease source. The latter factor is often the result of poor-quality treatment of fields with some herbicides. These examples prove the importance of sustainable agriculture not only for protecting the environment but also for a profitable business. Even an unfavourable light regime can cause negative consequences. Toxins that some embryophytes (higher plants) and fungi release into the soil can also be causal agents of crop diseases.

Infections' causal agents include

1. Bacteria
2. Viruses
3. Fungi
4. Nematodes
5. Parasitic plants.

Crop Diseases Caused By Bacteria

Among the most common infections in agriculture are crop diseases caused by bacteria. In this regard, the prevention and control of this kind of disease are pretty tricky. To infect the causal agent needs to get into the culture's tissue. It occurs mainly through damaged areas, such as caused by agricultural tools, insects (fleas), or simply unfavourable weather conditions (dust, wind, heavy rain). But bacteria can also infect plants through natural holes or glands (for example, which secrete nectar). Another feature of bacterial crop diseases is that causal agents, once in a plant or soil, can remain dormant for a long time until unfavourable conditions arise for them. First of all, significant temperature fluctuations and high levels of humidity act as catalysts for bacterial activity.

Symptoms of Bacterial Crop Diseases

The main bacterial disease indications include vascular wilting, necrosis, soft rot and tumour. Although this type of plant disease can be identified due to its pronounced symptoms, identifying a specific causal agent requires laboratory methods.

Common Bacterial Diseases

As noted earlier, due to a huge number of bacteria, there are many disease types. Here are some examples of the most common diseases of crop plants:

1. Black rot - Xanthomonas campestris
2. Bacterial canker - Clavibacter michiganensis
3. Bacterial soft rot - Pseudomonas spp
4. Bacterial leaf spot/Bacterial spot - Pseudomonas syringae - various strains
5. Bacterial wilt - seudomonas syringae pv. pisi
6. Bacterial leaf spot/Bacterial spot/Bacterial blight - Pseudomonas syringae
7. Bacterial brown spot - Pseudomonas syringae

Control Measures

Spraying of antibacterial solutions like Sour buttermilk etc.

Crop Diseases Caused By Fungi

Pathogenic fungi are the most typical agricultural problem. According to research, this plant disease type destroys about a third of all food crops every year. In this regard, this problem is severe both from a humanitarian and economic point of view. Like bacterial crop diseases, these infections affect plants mainly through wounds, stomata, and water pores. Also, fungal spores are often carried by gusts of wind.

Symptoms of Fungal Crop Diseases

Often, a fungal infection is expressed in local or general necrosis. Also, crop diseases caused by fungi can interfere with the average growth or contribute to its abnormal burst, called hypertrophy.

Common Fungal Diseases

1. White blister/White rust (Albugo candida)
2. Downy mildews (individual species damage particular crop families)
3. Powdery mildews (some species are restricted to particular crops or crop families)
4. Clubroot (Plasmodiophora brassicae)
5. Pythium species
6. Sclerotinia rots (S. sclerotiorum and S. minor) – a range of common names are used
7. Sclerotium rots (Sclerotium rolfsii and S. cepivorum)

Control Measures

Seed selection and treatment

Spraying of plant based fungicides

Crop Diseases Caused By Nematodes

Nematodes are parasitizing roundworms, which usually cannot be seen without special equipment. They live in the soil, and therefore mainly affect roots, tubers, and bulbs. There are over 4100 dangerous nematode species.

Symptoms of Nematode Crop Diseases

Essentially, nematodes feed by sucking juices from plants. Because of this, plants affected by these parasites often appear dried out, as if they are suffering from drought. Other symptoms are also similar: Yellowing, growth retardation, lack of response to fertilizers and water, the gradual general decline of a plant, reduction or even destruction of root systems. Although nematodes need a long period to cause significant damage to a plant, they spread exceptionally rapidly. Therefore, farmers should identify crop disease in the field to save the affected plants timely and prevent the disease from spreading.

Common Nematode Diseases

Diseases directly depend on the type of nematodes: Fusiform thickening of the stems is provoked by stem worms. The disease is expressed in the deformation of leaves, swelling of petioles, and the appearance of dark spots on tubers. Aphelenchoides, a disease of rice, is provoked by Aphelenchoides besseyi. Disease marks are blemishes on the tops of leaves, lack of grains, and culture depletion. The corresponding nematode Tylenchulus semipenetrans cause serious citrus diseases. It is characterized by the gradual death of not only leaves but also branches.

Crop Diseases Caused By Virus

The most minor but most critical plant enemies are viruses and viroids (subviral contagious agents). After infection, it is almost impossible to save a plant. Therefore the effect of plant diseases on crop production is of critical importance throughout the world. In most cases, the infection spreads as a result of healthy plants with sick contact. Viruses can also spread through vegetative reproduction, through seeds, pollen, and insects. But viruses most often spread through the soil.

Symptoms of Viral Crop Diseases

The symptoms of crop diseases caused by virus are usually divided into four types: malformations, such as abnormal growth of shoots, distortion of leaves and flowers; necrosis, wilting and the appearance of annular stripes and spots; dwarfism, growth retardation of both individual parts and the whole plant; discoloration, e.g. yellowing, and vein clearing. Root crop diseases, which are expressed in their rotting, are a characteristic indication of the presence of a virus. However, some plants may not show symptoms and are latent carriers of a disease. Therefore, extreme vigilance is required in the fight against this infection type.

Common Viral Diseases

Among the common examples of viral diseases in crop plants: Tobacco mosaic manifests in dwarfism and mosaic-like patterns on leaves. The disease is widespread throughout the world and can have significant economic consequences. Tomato spotted wilt is accompanied by the appearance and growth of necrotic yellow rings that gradually turn reddish-brown. Potato spindle tuber inhibits plant growth, tubers become fusiform and shrink.

Do soil microorganisms protect against crop disease?

They do. Soil microorganisms increase immunity, protect plants from many pathogens. Moreover, if beneficial microbes are present in the roots of a plant, it can fight pests more effectively while maintaining natural growth.

Control of Crop Diseases Caused By Viruses

Control the vectors :

Spraying of Cow urine based solutions to control

Crop Diseases Caused By Parasitic Plants

Parasitic plants are among the most dangerous plant pests in the world. With the help of particular organs, these plants settle in a host plant and satisfy at the expense of it (most often its vascular system). Although many parasites only weaken their "prey", some can kill a plant and pose a severe economic threat to agriculture. Depending on the species, parasites can attach from one to several dozen plant species.

Common Parasitic Plants

Today, there are about 400 parasitic plant species that have a substantial impact on the ecosystem in which they exist.

Few common examples

Mistletoe (Viscaceae): This semi-parasitic plant is widely represented throughout the world and is evergreen. As a parasite, mistletoe can exist on a significant number of plant species. It spreads thanks to special seeds that stick to birds and are carried with them to other plants. These seeds germinate through the host plant bark and connect to its food system. Cuscuta spp.: **Cuscuta spp.** : is a parasitic bindweed plant, which is very dangerous for various plants. It actively spreads, disrupts the metabolism of host plants, reduces productivity, and often leads to the death of plants. Moreover, Cuscuta spp. can be a carrier of viral plant and animal diseases. Because of all these features, the weed is a quarantine object.

Orobanche spp.: It is a dangerous root parasite without green leaves that cannot photosynthesize, and is utterly dependent on a host. For seed germination of Orobanche spp., it is required that a suitable plant is planted in the soil. Then the weed will attach to its roots and begin to receive ready-made food. The parasite leaves up to 100, 000 seeds. They remain viable in the ground for over ten years while waiting for a host. All it makes Orobanche spp. a dangerous pest.

Striga spp.: This group of parasitic plants is mainly found in tropical and subtropical regions and belongs to quarantine plants. In African countries, they are classified as a pandemic since Striga spp. can destroy up to 100 percent of a plant and cause irreparable economic damage. Primarily, this herb infects cereals but can also parasitize other plants, for example, when growing sugar cane. It's spread through seeds, growing together when ripe with the root system of a host plant. These parasites are very tenacious, so re-planting of a previously infected area is possible after nine years. In regions dependent on agriculture, the Striga spp. can even lead to the migration of people.

References

Aggarwal, P.; Viswamohanan, A.; Sharma, S. Unpacking India's Electricity Subsidies. International Institute for Sustainable Development. 2020. Available online: https://www.iisd.org/system/files/2020-12/india-electricity-subsidies.pdf (accessed on 2 August 2022).

Babalad, H.B.; Navali, G.V. Comparative Economics of Zero Budget Natural Farming with Conventional Farming Systems in Northern Dry Zone (Zone-3) of Karnataka. Econ. Aff. 2021, 66, 355–361.

Badagley C, Moghtader J, Quintero E, Zakem E, Chappell MJ, Avilés-Vázquez K, Samulon A, Perfecto I (2007). Organic agriculture and the global food supply. Renew. Agric. Food Syst., 22: 86-108.

Barbieri P, Pellerin S, Seufert V and Nesme T (2019) Changes in crop rotations would impact food production in an organically farmed world. Nature Sustainability, 2: 378-385.

Baron, G.L.; Jansen, V.A.; Brown, M.J.; Raine, N.E. Pesticide reduces bumblebee colony initiation and increases probability of population extinction. Nat. Ecol. Evol. 2017, 1, 1308–1316.

Bharucha, Z.P.; Mitjans, S.B.; Pretty, J. Towards redesign at scale through zero budget natural farming in Andhra Pradesh, India. Int. J. Agric. Sustain. 2020, 18, 1–20.

Cacho, M.M.T.G.; Giraldo, O.F.; Aldasoro, M.; Morales, H.; Ferguson, B.G.; Rosset, P.; Khadse, A.; Campos, C. Bringing agroecology to scale: Key drivers and emblematic cases. Agroecol. Sustain. Food Syst. 2018, 42, 637.

Canfield D, Glazer A and Falkowski P (2010). The evolution and future of earth's nitrogen cycle. Science 330, 192–196.

Connor DJ (2008). Organic agriculture cannot feed the world. Field Crops Research, 106 (2): 187-190.

Dasgupta, B. India's green revolution. Econ. Polit. Wkly. 1977, 12, 241–260.

Duddigan, S.; Collins, C.D.; Hussain, Z.; Osbahr, H.; Shaw, L.J.; Sinclair, F.; Sizmur, T.; Thallam, V.; Winowiecki, L.A. Impact of Zero Budget Natural Farming on Crop Yields in Andhra Pradesh, SE India. Sustainability 2022, 14, 1689.

Economic Survey (2019). Economic Survey 2018-19. Department of Economic Affairs, Ministry of Finance, Government of India.

Economic Survey, Agriculture and Food Management. Available online: https://www.indiabudget.gov.in/economicsurvey/ (accessed on 2 August 2022).

EPW (2019). Mirage of Zero Budget Farming (Editorials). Economic & Political Weekly, Vol. LIV (30), July 27.

Evenson RE and Gollin D (2003) Assessing the impact of the green revolution, 1960– 2000. Science 300, 758–762.

Fertilizer Association of India. Available online: https://www.faidelhi.org/general/subsidy-fert.pdf (accessed on 25 September 2022).

Fukuoka M. (1987). The Natural Way of Farming: The Theory and Practice of Green Philosophy. Japan Publications. ISBN 978-0-87040-613-3.

Gilbert SF, Sapp J and Tauber AI. (2012). A symbiotic view of life: we have never been individuals. Quarterly Review of Biology 87(4): 325–341.

Hazell P. and Rahman A. (2014). New Directions for Smallholder Agriculture. Oxford University Press.

IAASTD (2009). Agriculture at a Crossroads: The Global Report. Island Press, Washington DC, USA.

Jain, S. Natural farming: Is India ready to bring 14 million hectares land under organic agriculture? Firstpost, 7 November 2022.

John, D.A.; Babu, G.R. Lessons from the aftermaths of green revolution on food system and health. Front. Sustain. Food Syst. 2021, 5, 21.

Khadse, A.; Rosset, P.M.; Morales, H.; Ferguson, B.G. Taking agroecology to scale: The Zero Budget Natural Farming peasant movement in Karnataka, India. J. Peasant Stud. 2018, 45, 192.

Khadse, A.; Rosset, P.M. Zero Budget Natural Farming in India—From inception to institutionalization. Agroecol. Sustain. Food Syst. 2019, 43, 848–871.

Khadse A, Rosset PM, and Ferguson BG. (2017). Taking agro ecology to scale: The zero budget natural farming peasant movement in Karnataka, India. The Journal of Peasant Studies 45:1–28.

Koner, N.; Laha, A. Economics of alternative models of organic farming: Empirical evidences from zero budget natural farming and scientific organic farming in West Bengal, India. Int. J. Agric. Sustain. 2021, 19, 255–268.

Kumar, S.; Kale, P.; Thombare, P. Zero Budget Natural Farming (ZBNF): Securing smallholder farming from distress. Sci. Agric. Allied Sect. 2019, 1, 3.

Kumar, V. A Question of Sales: Natural Farming Faces Challenges in Himachal; Here Is How, Downtoearth. Available online: https://www.downtoearth.org.in/news/agriculture/a-question-of-sales-natural-farming-faces-challenges-in-himachalhere-is-how-84699 (accessed on 15 February 2023).

Kumar Vijay T, Raidu DV, Killi J, Pillai M, Shah P, Kalavakonda V and Lakhey S. (2009). Ecologically Sound, Economically Viable Community Managed Sustainable Agriculture in Andhra Pradesh, India. The World Bank, Washington DC.

Lorimer J. (2017). Probiotic environmentalities: rewilding with wolves and worms. Theory, Culture and Society, 34(4): 27–48.

Ministry of Agriculture and Farmers' Welfare. Agricultural Statistics at a Glance-2021; Directorate of Economics and Statistics, Government of India: Delhi, India, 2022.

Mishra, S. Zero Budget Natural Farming: Are This and Similar Practices the Answers; Nabakrushna Choudhury Centre for Development Studies (NCDS): Odisha, India, 2018.

Mishra S. (2018). Zero Budget Natural Farming: Are this and similar practices the answers? Working Paper No. 70, Nabakrushna Choudhury Centre for Development Studies, Bhubaneswar.

MoA&FW (2019). Agriculture Census 2015-16 (Phase-I). All India Report on Number and Area of Operational Holdings. Ministry of Agriculture & Farmers Welfare, Government of India.

Munster D. (2018). Performing alternative agriculture: Critique and recuperation in Zero Budget Natural Farming, South India. Journal of Political Ecology, 25: 748- 764.

Murthy, R. L. N. and Venateswarulu, P., 1998. Introducing eco-friendly farming techniques and inputs in cotton. In: Proceedings of the workshop on 'Eco-friendly cotton, 1998' held at Agricultural College & Research Institute, Madurai, Tamil Nadu, Oct. 27-28, 1995 p.110.

Naik, A.K.; Brunda, S.; Chaithra, G.M. Comparative Economic Analysis of Zero Budget Natural Farming for Kharif Groundnut under Central Dry Zone of Karnataka, India. J. Econ. Manag. Trade 2020, 26, 27–34.

National Rainfed Area Authority. Report on Crop Feasibility Study to Recommend Appropriate Mechanisms for Providing Farmers with Rational Compensation on Occurrence of Crop Losses and Identifying Vulnerable Districts for Risk Coverage under Pradhan Mantri Fasal Bima Yojna (PMFBY). 2022. Available online: https://pmfby.gov.in/compendium/General/1_2_3_merged. pdf (accessed on 30 November 2022).

Niyogi DG. (2018). Andhra farmers taste success with Zero Budget Natural Farming. Down to Earth, (Online).

Niyogi DG. (2018). Andhra farmers taste success with Zero Budget Natural Farming. Down to Earth, (Online). https://www.downtoearth.org.in/author/deepanwitagita-niyogi-2399.

Palekar S. (2005). The philosophy of spiritual farming. 2nd ed. Zero Budget Natural Farming Research, Development & Extension Movement, Amravati, Maharashtra, India.

Pingali PL. (2012). Green revolution: impacts, limits, and the path ahead. Proceedings of the National Academy of Sciences of the United States of America, 109(31):12302– 12308. doi:10.1073/pnas.0912953109.

Press Information Bureau. Pilot Study on Zero Budget Natural Farming Initiated at 4 Locations: Shri Narendra Singh Tomar. 2019. Available online: https://pib.gov.in/PressReleaseIframePage.aspx?PRID=1882245 (accessed on 15 February 2023).

Press Information Bureau: PM Speech at National Conclave on Natural Farming. Available online: https://www.pib.gov.in/ PressReleseDetail.aspx?PMO=3&PRID=1782250 (accessed on 5 October 2022).

Rosset, P.M.; Martínez-Torres, M.E. Rural social movements and agroecology: Context, theory, and process. Ecol. Soc. 2012, 17, 17.

Saharan, B.S.; Tyagi, S.; Kumar, R.; Vijay; Om, H.; Mandal, B.S.; Duhan, J.S. Application of Jeevamrit Improves Soil Properties in Zero Budget Natural Farming Fields. Agriculture 2023, 13, 196.

Saldanha LF. (2018). A Review of Andhra Pradesh's Climate Resilient Zero Budget Natural Farming, Environment Support Group, Bangalore.

Searchinger TD, Wirsenius S, Beringer T and Dumas P. (2018). Assessing the efficiency of changes in land use for mitigating climate change. Nature, 564: 249-253.

Shotwell A. (2016). Against Purity: Living Ethically in Compromised Times. Minnepolis, MN: University of Minnesota Press.

Shyam, D.M.; Sreenath, D.; Rajesh, N.; Gajanan, S.; Girish, C. Zero budget natural farming-an empirical analysis. Green Farming 2019, 106, 661–667.

Smith, J.; Yeluripati, J.; Smith, P.; Nayak, D.R. Potential yield challenges to scale-up of zero budget natural farming. Nat. Sustain. 2020, 3, 247–252.

Smith P et al. (2013) How much land based greenhouse gas mitigation can be achieved without compromising food security and environmental goals? Global Change Biology, 19 (8). https://doi.org/10.1111/gcb.12160.

The Fertiliser Association of India: 67th Annual Report 2021–2022. Available online: https://www.faidelhi.org/general/FAI-AR21-22.pdf (accessed on 2 December 2022).

Wallenstein M. (2017). To restore our soils, feed the microbes. The Conversation. Online article, visited on 22 Dec 2019.

Index